CODES FOR ERROR DETECTION

Series on Coding Theory and Cryptology

Editors: Harald Niederreiter *(National University of Singapore, Singapore)* and
San Ling *(Nanyang Technological University, Singapore)*

Published

Series on Coding Theory and Cryptology – Vol. 2

CODES FOR ERROR DETECTION

Torleiv Kløve

University of Bergen, Norway

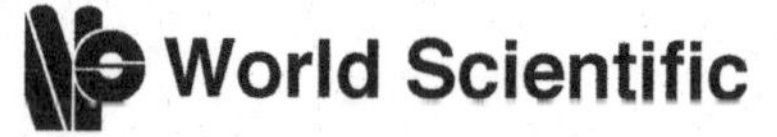

NEW JERSEY • LONDON • SINGAPORE • BEIJING • SHANGHAI • HONG KONG • TAIPEI • CHENNAI

Published by

World Scientific Publishing Co. Pte. Ltd.

5 Toh Tuck Link, Singapore 596224

USA office: 27 Warren Street, Suite 401-402, Hackensack, NJ 07601

UK office: 57 Shelton Street, Covent Garden, London WC2H 9HE

British Library Cataloguing-in-Publication Data
A catalogue record for this book is available from the British Library.

Series on Coding Theory and Cryptology — Vol. 2
CODES FOR ERROR DETECTION

Copyright © 2007 by World Scientific Publishing Co. Pte. Ltd.

All rights reserved. This book, or parts thereof, may not be reproduced in any form or by any means, electronic or mechanical, including photocopying, recording or any information storage and retrieval system now known or to be invented, without written permission from the Publisher.

For photocopying of material in this volume, please pay a copying fee through the Copyright Clearance Center, Inc., 222 Rosewood Drive, Danvers, MA 01923, USA. In this case permission to photocopy is not required from the publisher.

ISBN-13 978-981-270-586-0
ISBN-10 981-270-586-4

Printed in Singapore.

S.D.G.

Preface

There are two basic methods of error control for communication, both involving coding of the messages. The differences lay in the way the codes are utilized. The codes used are block codes, which are the ones treated in this book, or convolutional codes. Often the block codes used are linear codes. With forward error correction, the codes are used to detect and correct errors. In a repeat request system, the codes are used to detect errors, and, if there are errors, request a retransmission.

Usually it is a much more complex task to correct errors than merely detect them. Detecting errors has the same complexity as encoding, which is usually linear in the length of the codewords. Optimal error correcting decoding is in general an NP-hard problem, and efficient decoding algorithms are known only for some classes of codes. This has generated much research into finding new classes of codes with efficient decoding as well as new decoding algorithms for known classes of codes.

There are a number of books on error control, some are listed in the bibliography at the end of this book. The main theme of almost all these books is error correcting codes. Error detection tends to be looked upon as trivial and is covered in a few pages at most. What is then the reason behind the following book which is totally devoted to error detecting codes? The reason is, on the one hand, that error detection combined with repeat requests is a widely used method of error control, and on the other hand, that the analysis of the reliability of the information transmission system with error detection is usually quite complex. Moreover, the methods of analysis are often not sufficiently known, and simple rules of the thumb are used instead.

The main parameter of a code for error detection is its *probability of error detection*, and this is the main theme of this book. There are many papers devoted to the study of the probability of undetected error, both for symmetric channels and other channels, with or without memory. They are spread over many journals and conference proceedings. In Kløve and Korzhik (1995), which was published twelve years ago, we collected the results then known, mainly for linear codes on the binary symmetric channel, and presented them in a unified form. We also included a number of new results. In the last twelve years a number of significant new results have been published (in approximately one hundred new papers). In the present book, we have included all the important new results, and we also include some new unpublished results. As far as possible, the results are given for both linear and non-linear codes, and for general alphabet size. Many results previously published for binary codes or for linear codes (or both) are generalized.

We have mainly restricted ourselves to channels without memory and to codes for error detection only (not combined error correction and detection since these results belong mainly with error correcting codes; the topic is briefly mentioned, however).

Chapter 1 is a short introduction to coding theory, concentrating on topics that are relevant for error detection. In particular, we give a more detailed presentation of the distance distribution of codes than what is common in books on error correction.

Chapter 2 is the largest chapter, and it contains a detailed account of the known results on the probability of undetected error on the q-ary symmetric channel. Combined detection and correction will be briefly mentioned from the error detection point of view.

Chapter 3 presents results that are particular for the binary symmetric channel.

Chapter 4 considers codes for some other channels.

Each chapter includes a list of comments and references.

Finally, we give a bibliography of papers on error detection and related topics.

The required background for the reader of this book will be some basic knowledge of coding theory and some basic mathematical knowledge: algebra (matrices, groups, finite fields, vector spaces, polynomials) and elementary probability theory.

Bergen, January 2007
T. Kløve

Contents

Chapter 1

Basics on error control

1.1 ABC on codes

1.1.1 *Basic notations and terminology*

The basic idea of coding is to introduce redundancy that can be utilized to detect and, for some applications, correct errors that have occurred during a transmission over a channel. Here "transmission" is used in a wide sense, including any process which may corrupt the data, e.g. transmission, storage, etc. The symbols transmitted are from some finite alphabet F. If the alphabet has size q we will sometimes denote it by F_q. We mainly consider channels without memory, that is, a symbol $a \in F$ is transformed to $b \in F$ with some probability $\pi(a, b)$, independent of other symbols transmitted (earlier or later). Since the channel is described by the transition probabilities and a change of alphabet is just a renaming of the symbols, the actual alphabet is not important. However, many code constructions utilize a structure of the alphabet. We will usually assume that the alphabet of size q is the set Z_q of integers modulo q. When q is a prime power, we will sometimes use the finite field $GF(q)$ as alphabet. The main reason is that vector spaces over finite fields are important codes; they are called linear codes.

As usual, F^n denotes the set of n-tuples $(a_1, a_2, \cdots, a_n)$ where $a_i \in F$. The n-tuples will also be called *vectors.*

Suppose that we have a set $\mathcal{M}$ of M possible messages that may be sent. An $(n, M; q)$ *code* is a subset of F^n containing M vectors. An *encoding* is a one-to-one function from $\mathcal{M}$ to the code. The vectors of the code are called *code words.*

1.1.2 *Hamming weight and distance*

The *Hamming weight* $w_{\rm H}(\mathbf{x})$ of a vector $\mathbf{x}$ is the number of non-zero positions in $\mathbf{x}$, that is

$$w_{\rm H}(\mathbf{x}) = \#\{i \mid 1 \le i \le n \text{ and } x_i \ne 0\}.$$

The *Hamming distance* $d_{\rm H}(\mathbf{x},\mathbf{y})$ between two vectors $\mathbf{x},\mathbf{y} \in F_q^n$ is the number of positions where they differ, that is

$$d_{\rm H}(\mathbf{x},\mathbf{y}) = \#\{i \mid 1 \le i \le n \text{ and } x_i \ne y_i\}.$$

If a vector $\mathbf{x}$ was transmitted and e errors occurred during transmission, then the received vector $\mathbf{y}$ differs from $\mathbf{x}$ in e positions, that is $d_{\rm H}(\mathbf{x},\mathbf{y}) = e$.

Clearly,

$$d_{\rm H}(\mathbf{x},\mathbf{y}) = w_{\rm H}(\mathbf{x}-\mathbf{y}).$$

For an $(n, M; q)$ code C, define the *minimum distance* by

$$d = d(C) = \min\{d_{\rm H}(\mathbf{x},\mathbf{y}) \mid \mathbf{x},\mathbf{y} \in C, \quad \mathbf{x} \ne \mathbf{y}\},$$

and let

$$d(n, M; q) = \max\{d(C) \mid C \text{ is an } (n, M; q) \text{ code}\}.$$

Sometimes we include $d = d(C)$ in the notation for a code and write $(n, M, d; q)$ and $[n, k, d; q]$. The *rate* of a code $C \subset F_q^n$ is

$$R = \frac{\log_q \#C}{n}.$$

Define

$$\delta(n, R; q) = \frac{d\left(n, \lceil q^{Rn} \rceil ; q\right)}{n},$$

and

$$\delta(R; q) = \limsup_{n\to\infty} \delta(n, R; q).$$

1.1.3 *Support of a set of vectors*

For $\mathbf{x} = (x_1, x_2, \ldots, x_n), \mathbf{y} = (y_1, y_2, \ldots, y_n) \in F^n$, we define the *support* of the pair $(\mathbf{x},\mathbf{y})$ by

$$\chi(\mathbf{x},\mathbf{y}) = \{i \in \{1, 2, \ldots, n\} \mid x_i \ne y_i\}.$$

Note that

$$\#\chi(\mathbf{x},\mathbf{y}) = d_{\rm H}(\mathbf{x},\mathbf{y}).$$

For a single vector $\mathbf{x} \in F^n$, the support is defined by $\chi(\mathbf{x}) = \chi(\mathbf{x},\mathbf{0})$.

For a set $S \subseteq F^n$, we define its *support* by

$$\chi(S) = \bigcup_{\mathbf{x},\mathbf{y}\in S} \chi(\mathbf{x},\mathbf{y}).$$

In particular, $\chi(S)$ is the set of positions where not all vectors in S are equal.

1.1.4 *Extending vectors*

Let $\mathbf{x} = (x_1, x_2, \cdots, x_n) \in F^n$, $\mathbf{y} = (y_1, y_2, \cdots, y_m) \in F^m$ and $u \in F$. Then

$$u\mathbf{x} = (ux_1, ux_2, \cdots, ux_n),$$

$$(\mathbf{x}|u) = (x_1, x_2, \cdots, x_n, u),$$

$$(\mathbf{x}|\mathbf{y}) = (x_1, x_2, \cdots, x_n, y_1, y_2, \cdots, y_m).$$

The last two operations are called concatenation. For a subset S of F^n,

$$uS = \{u\mathbf{x} \mid \mathbf{x} \in S\},$$

$$(S|u) = \{(\mathbf{x}|u) \mid \mathbf{x} \in S\}.$$

1.1.5 *Ordering*

Let some ordering $\leq$ of F be given. We extend this to a partial ordering of F^n as follows:

$$(x_1, x_2, \cdots, x_n) \leq (y_1, y_2, \cdots, y_n) \text{ if } x_i \leq y_i \text{ for } 1 \leq i \leq n.$$

For Z_q we use the natural ordering $0 < 1 < 2 \cdots < q - 1$.

1.1.6 *Entropy*

The base q (or q-ary) *entropy function* $H_q(z)$ is defined by

$$H_q(z) = -z \log_q\Big(\frac{z}{q-1}\Big) - (1-z)\log_q(1-z)$$

for $0 \leq z \leq 1$. $H_q(z)$ is an increasing function on $\left[0, \frac{q-1}{q}\right]$, $H_q(0) = 0$, and $H_q\left(\frac{q-1}{q}\right) = 1$. Define $\rho(z) = \rho_q(z)$ on $[0,1]$ by $\rho_b(z) \in \left[0, \frac{q-1}{q}\right]$ and

$$H_b(\rho_b(z)) = 1 - z.$$

1.1.7 *Systematic codes*

An $(n, q^k; q)$ code C is called *systematic* if it has the form

$$C = \{(\mathbf{x}|f(\mathbf{x})) \mid \mathbf{x} \in F_q^k\}$$

where f is a mapping from F_q^k to F_q^{n-k}. Here $(\mathbf{x}|f(\mathbf{x}))$ denotes the concatenation of $\mathbf{x}$ and $f(\mathbf{x})$.

1.1.8 *Equivalent codes*

Two $(n, M; q)$ codes C_1, C_2 are *equivalent* if C_2 can be obtained from C_1 by permuting the positions of all code words by the same permutation. We note that equivalent codes have the same distance distribution, and in particular the same minimum distance.

1.1.9 *New codes from old*

There are a number of ways to construct new codes from one or more old ones. We will describe some of these briefly. In a later section we will discuss how the error detecting capability of the new codes are related to the error detecting capability of the old ones.

Extending a code

Consider an $(n, M; q)$ code C. Let $\mathbf{b} = (b_1, b_2, \cdots, b_n) \in F_q^n$. Let C^{ex} be the $(n+1, M; q)$ code

$$C^{\mathrm{ex}} = \Big\{(a_1, a_2, \cdots, a_n, -\sum_{i=1}^{n} a_i b_i) \;\Big|\; (a_1, a_2, \cdots, a_n) \in C\Big\}.$$

Note that this construction depends on the algebraic structure of the alphabet F_p (addition and multiplication are used to define the last term). For example, let $n = 2$, $\mathbf{b} = (1, 1)$, and $\mathbf{a} = (1, 1)$. If the alphabet is $GF(4)$, then $a_1 b_1 + a_2 b_2 = 0$, but if the alphabet is Z_4, then $a_1 b_1 + a_2 b_2 = 2 \neq 0$.

Puncturing a code

Consider an $(n, M; q)$ code. *Puncturing* is to remove the first position from each code word (puncturing can also be done in any other position). This produces a code C^{p} of length $n-1$. If two code words in C are identical, except in the first position, then the punctured code words are the same. Hence the size of C^{p} may be less than M. On the other hand, any code word $\mathbf{c} \in C^{\mathrm{p}}$ is obtained from a vector $(a|\mathbf{c})$ where $a \in F_q$. Hence, the size of C^{p} is at least M/q. The minimum distance may decrease by one. Clearly, the operation of puncturing may be repeated.

Shortening a code

Consider an $(n, M; q)$ code C with the first position in its support. *Shortening* (by the first position) we obtain the $(n-1, M'; q)$ code

$$C^{\mathrm{s}} = \left\{ \mathbf{x} \in F^{n-1} \;\middle|\; (0|\mathbf{x}) \in C \right\},$$

that is, we take the set of all code words of C with 0 in the first position and remove that position. More general, we can shorten by any position in the support of the code.

We note that shortening will not decrease the minimum distance; however it may increase it. In the extreme case, when there are no code words in C with 0 in the first position, C^{s} is empty.

Zero-sum subcodes of a code

Consider an $(n, M; q)$ code C. The *zero-sum subcode* C^{zs} is the code

$$C^{\mathrm{zs}} = \left\{ (a_1, a_2, \cdots, a_n) \in C \;\middle|\; \sum_{i=1}^{n} a_i = 0 \right\}.$$

Also this construction depends on the algebraic structure of the alphabet.

In the binary case, $\sum_{i=1}^{n} a_i = 0$ if and only if $w_{\mathrm{H}}(\mathbf{a})$ is even, and C^{zs} is then called the *even-weight subcode.*

1.1.10 *Cyclic codes*

A code $C \subseteq F^n$ is called *cyclic* if

$$(a_{n-1}, a_{n-2}, \cdots, a_0) \in C \text{ implies that } (a_{n-2}, a_{n-3}, \cdots, a_0, a_{n-1}) \in C.$$

Our reason for the special way of indexing the elements is that we want to associate a polynomial in the variable z with each n-tuple as follows:

$$\mathbf{a} = (a_{n-1}, a_{n-2}, \ldots, a_0) \leftrightarrow a(z) = a_{n-1}z^{n-1} + a_{n-2}z^{n-2} \cdots + a_0.$$

This correspondence has the following property (it is an isomorphism): if $\mathbf{a}, \mathbf{b} \in F^n$ and $c \in F$, then

$$\mathbf{a} + \mathbf{b} \leftrightarrow a(z) + b(z),$$
$$c\mathbf{a} \leftrightarrow ca(z).$$

In particular, any code may be represented as a set of polynomials. Moreover, the polynomial corresponding to $(a_{n-2}, a_{n-3}, \cdots, a_0, a_{n-1})$ is

$$a_{n-1} + \sum_{i=0}^{n-2} a_i z^{i+1} = za(z) - a_{n-1}(z^n - 1) \equiv za(z) \pmod{z^n - 1}.$$

1.2 Linear codes

An $[n,k;q]$ *linear* code is a k-dimensional subspace of $GF(q)^n$. This is in particular an $(n,q^k;q)$ code. Vector spaces can be represented in various ways and different representations are used in different situations.

1.2.1 *Generator and check matrices for linear codes*

Suppose that $\{\mathbf{g}_1,\mathbf{g}_2,\cdots,\mathbf{g}_k\}$ is a basis for C. Then C is the set of all possible linear combinations of these vectors. Let G be the $k \times n$ matrix whose k rows are $\mathbf{g}_1,\mathbf{g}_2,\cdots,\mathbf{g}_k$. Then

$$C = \{\mathbf{x}G \mid \mathbf{x} \in GF(q)^k\}.$$

We call G a *generator matrix* for C. A natural *encoding* $GF(q)^k \to GF(q)^n$ is given by

$$\mathbf{x} \mapsto \mathbf{x}G.$$

If $T: GF(q)^k \to GF(q)^k$ is a linear invertible transformation, then TG is also a generator matrix. The effect is just a change of basis.

The *inner product* of two vectors $\mathbf{x},\mathbf{y} \in GF(q)^n$ is defined by

$$\mathbf{x}\cdot\mathbf{y} = \mathbf{x}\mathbf{y}^{\mathrm{t}} = \sum_{i=1}^{n} x_i y_i,$$

where $\mathbf{y}^{\mathrm{t}}$ is the transposed of $\mathbf{y}$. For a linear $[n,k;q]$, the dual code is the $[n,n-k;q]$ code

$$C^{\perp} = \{\mathbf{x} \in GF(q)^n \mid \mathbf{x}\mathbf{c}^{\mathrm{t}} = 0 \text{ for all } \mathbf{c} \in C\}.$$

If H is a generator matrix for $C^{\perp}$, then

$$C = \{\mathbf{x} \in GF(q)^n \mid \mathbf{x}H^t = 0\},$$

where H^t is the transposed of H. H is known as a *(parity) check matrix* for C. Note that $GH^{\mathrm{t}} = 0$ and that any $(n-k)\times n$ matrix H of rank $n-k$ such that $GH^{\mathrm{t}} = 0$ is a check matrix.

1.2.2 *The simplex codes and the Hamming codes*

Before we go on, we define two classes of codes, partly because they are important in their own right, partly because they are used in other constructions.

Let Γ_k be a $k \times \frac{q^k-1}{q-1}$ matrix over $GF(q)$ such that

(i) all columns of Γ_k are non-zero,

(ii) if $\mathbf{x} \neq \mathbf{y}$ are columns, then $\mathbf{x} \neq j\mathbf{y}$ for all $j \in GF(q)$.

The matrix Γ_k generates a $\left[\frac{q^k-1}{q-1}, k, q^{k-1}; q\right]$ code S_k whose non-zero code words all have weight q^{k-1}. It is known as the *Simplex code*. The dual code is an $\left[\frac{q^k-1}{q-1}, \frac{q^k-1}{q-1} - k, 3; q\right]$ code known as the *Hamming code.*

1.2.3 *Equivalent and systematic linear codes*

Let C_1 be an $[n, k; q]$ code and let

$$C_2 = \{\mathbf{x}Q\Pi \mid \mathbf{x} \in C_1\}$$

where Q is a non-singular diagonal $n \times n$ matrix and Π is an $n \times n$ permutation matrix. If G is a generator matrix for C_1, then $GQ\Pi$ is a generator matrix for C_2.

Let G be a $k \times n$ generator matrix for some linear code C. By suitable row operations this can be brought into reduced echelon form. This matrix will generate the same code. A suitable permutation of the columns will give a matrix of the form $(I_k|P)$ which generates a systematic code. Here I_k is the identity matrix and P is some $k \times (n-k)$ matrix. Therefore, any linear code is equivalent to a systematic linear code. Since

$$(I_k|P)(-P^{\mathrm{t}}|I_{n-k})^{\mathrm{t}} = -P + P = 0,$$

$H = (-P^{\mathrm{t}}|I_{n-k})$ is a check matrix for C.

1.2.4 *New linear codes from old*

Extending a linear code

If C is a linear code, then C^{ex} is also linear. Moreover, if H is a check matrix for C, then a check matrix for C^{ex} (where C is extended by $\mathbf{b}$) is

$$\begin{pmatrix} H & \mathbf{0}^{\mathrm{t}} \\ \mathbf{b} & 1 \end{pmatrix}.$$

In particular, in the binary case, if $b_1 = b_2 = \cdots = b_n = 1$, we have extended the code with a parity check. The code $(GF(2)^n)^{\mathrm{ex}}$ is known as the *single parity check code* or just the parity check code.

Shortening a linear code

Shortening a linear code gives a new linear code. If $G = (I_k|P)$ generates a systematic linear code and the code is shortened by the first position, then a generator matrix for the shortened code is obtained by removing the first row and the first column of G.

Puncturing a linear code

Consider puncturing an $[n, k, d; q]$ code C. If the position punctured is not in the support of C, then $C^{\rm p}$ is an $[n-1, k, d; q]$ code. If the position punctured is in the support of C, then $C^{\rm p}$ is an $[n-1, k-1, d'; q]$ code. If $d > 1$, then $d' = d$ or $d' = d-1$. If $d = 1$, then d' can be arbitrary large. For example, if C is the $[n, 2, 1; 2]$ code generated by $(1, 0, 0, \ldots, 0)$ and $(1, 1, 1, \ldots, 1)$, and we puncture the first position, the resulting code is a $[n-1, 1, n-1; 2]$ code.

*The *-operation for linear codes*

Let C be an $[n, k; q]$ code over $GF(q)$. Let C^* denote the $\left[n + \frac{q^k-1}{q-1}, k; q\right]$ code obtained from C by extending each code word in C by a distinct code word from the simplex code S_k. We remark that the construction is not unique since there are many ways to choose the code words from S_k. However, for error detection they are equally good (we will return to this later).

We also consider iterations of the *-operation. We define C^{r*} by

$$C^{0*} = C,$$
$$C^{(r+1)*} = (C^{r*})^*.$$

Product codes

Let C_1 be an $[n_1, k_1, d_1; q]$ code and C_2 an $[n_2, k_2, d_2; q]$ code. The *product code* is the $[n_1 n_2, k_1 k_2, d_1 d_2; q]$ code C whose code words are usually written as an $n_1 \times n_2$ array; C is the set of all such arrays where all rows belong to C_1 and all columns to C_2.

Tensor product codes

Let C_1 be an $[n_1, k_1; q]$ code with parity check matrix

$$H_1 = (h_{ij}^{[1]})_{1 \le i \le n_1 - k_1, 1 \le j \le n_1}$$

and C_2 an $[n_2, k_2; q]$ code with parity check matrix

$$H_2 = (h_{ij}^{[2]})_{1 \le i \le n_2 - k_2, 1 \le j \le n_2}.$$

The tensor product code is the $[n_1 n_2, n_1 k_2 + n_2 k_1 - k_1 k_2; q]$ code with parity matrix $H = (h_{ij})$ which is the tensor product of H_1 and H_2, that is

$$h_{i_1(n_2-k_2)+i_2, j_1 n_2 + j_2} = h_{i_1, j_1}^{[1]} h_{i_2, j_2}^{[2]}.$$

Repeated codes

Let C be an $(n, M; q)$ code and let r be a positive integer. The r times repeated code, C^r is the code

$$C^r = \{(\mathbf{c}_1|\mathbf{c}_2|\cdots|\mathbf{c}_r) \mid \mathbf{c}_1, \mathbf{c}_2, \ldots, \mathbf{c}_r \in C\},$$

that is, the Cartesian product of r copies of C. This is an $(rn, M^r; q)$ code with the same minimum distance as C.

Concatenated codes

Codes can be concatenated in various ways. One such construction that has been proposed for a combined error correction and detection is the following.

Let C_1 be an $[N, K; q]$ code and C_2 an $[n, k; q]$ code, where $N = mk$ for some integer m. The encoding is done as follows: K information symbols are encoded into N symbols using code C_1. These $N = mk$ are split into m blocks with k symbols in each block. Then each block is encoded into n symbols using code C_2. The concatenated code is an $[mn, K; q]$ code. If G_1 and G_2 are generator matrices for C_1 and C_2 respectively, then a generator matrix for the combined code is the following.

$$G_1 \begin{pmatrix} G_2 & 0 & \cdots & 0 \\ 0 & G_2 & \cdots & 0 \\ \vdots & \vdots & \ddots & \vdots \\ 0 & 0 & \cdots & G_2 \end{pmatrix}.$$

The construction above can be generalized in various ways. One generalization that is used in several practical systems combines a convolutional code for error correction and a block code (e.g. an CRC code) for error detection.

1.2.5 *Cyclic linear and shortened cyclic linear codes*

Many important codes are cyclic linear codes or shortened cyclic linear codes. One reason that cyclic codes are used is that they have more algebraic structure than linear codes in general, and this structure can be used both in the analysis of the codes and in the design of efficient encoders and decoders for error correction. For example, the roots of the polynomial $g(z)$, given by the theorem below, give information on the minimum distance of the code. Hamming codes is one class of cyclic codes and shortened Hamming codes and their cosets are used in several standards for data transmission where error detection is important. This is our main reason for introducing them in this text.

Theorem 1.1. *Let C be a cyclic $[n, k; q]$ code. Then there exists a monic polynomial $g(z)$ of degree $n - k$ such that*

$$C = \{v(z)g(z) \mid \deg(v(z)) < k\}.$$

Proof. Let $g(z)$ be the monic polynomial in C of smallest positive degree, say degree m. Then $z^i g(z) \in C$ for $0 \le i < n-m$. Let $a(z)$ be any non-zero polynomial in C, of degree s, say; $m \le s < n$. Then there exist elements $c_{s-m}, c_{s-m-1}, \cdots, c_0 \in GF(q)$ such that

$$r(z) = a(z) - \sum_{i=0}^{s-m} c_i z^i g(z)$$

has degree less than m (this can easily be shown by induction on s). Since C is a linear code, $r(z) \in C$. Moreover, there exists a $c \in GF(q)$ such that $cr(z)$ is monic, and the minimality of the degree of $g(z)$ implies that $r(z)$ is identically zero. Hence $a(z) = v(z)g(z)$ where $v(z) = \sum_{i=0}^{s-m} c_i z^i$. In particular, the set

$$\{g(z), zg(z), \cdots, z^{n-1-m}g(z)\}$$

of $n - m$ polynomials is a basis for C and so $n - m = k$, that is

$$k = n - m.$$

□

The polynomial $g(z)$ is called the *generator polynomial* of C.

If $g(1) \neq 0$, then the code generated by $(z-1)g(z)$ is an $[n+1, k; q]$ code. It is the code C^{ex} obtained from C extending using the vector $\mathbf{1} = (11 \cdots 1)$, that is

$$\left\{(a_1, a_2, \cdots, a_n, -\sum_{i=1}^{n} a_i) \;\middle|\; (a_1, a_2, \cdots, a_n) \in C\right\}.$$

Encoding using a cyclic code is usually done in one of two ways. Let $\mathbf{v} = (v_{k-1}, v_{k-2}, \cdots, v_0) \in GF(q)^k$ be the information to be encoded. The first, and direct way of encoding, is to encode into $v(z)g(z)$. On the other hand, the code is systematic, but this encoding is not. The other way of encoding is to encode $\mathbf{v}$ into the polynomial in C "closest" to $z^{n-k}v(z)$. More precisely, there is a unique $a(z)$ of degree less than k such that

$$-r(z) = z^{n-k}v(z) - a(z)g(z)$$

has degree less than $n - k$, and we encode into

$$a(z)g(z) = z^{n-k}v(z) + r(z).$$

The corresponding code word has the form $(\mathbf{v}|\mathbf{r})$, where $\mathbf{r} \in GF(q)^{n-k}$.

Theorem 1.2. *Let C be a cyclic $[n, k; q]$ code with generator polynomial $g(z)$. Then $g(z)$ divides $z^n - 1$, that is, there exists a monic polynomial $h(z)$ of degree k such that*

$$g(z)h(z) = z^n - 1.$$

Moreover, the polynomial

$$\tilde{h}(z) = -g(0)z^k h\left(\frac{1}{z}\right)$$

is the generator polynomial of $C^\perp$.

Proof. There exist unique polynomials $h(z)$ and $r(z)$ such that

$$z^n - 1 = g(z)h(z) + r(z)$$

and $\deg(r(z)) < n - k$. In particular $r(z) \equiv h(z)g(z) \pmod{z^n - 1}$ and so $r(z) \in C$. The minimality of the degree of $g(z)$ implies that $r(z) \equiv 0$.

Let $g(z) = \sum_{i=0}^{n-k} g_i z^i$ and $h(z) = \sum_{i=0}^{k} h_i z^i$. Then

$$\sum_i g_{l-i}h_i = \begin{cases} -1 \text{ if } & l = 0, \\ 0 \text{ if } & 0 < l < n, \\ 1 \text{ if } & l = n. \end{cases}$$

Further, $\tilde{h}(z) = -g_0 \sum_{i=0}^{k} h_{k-i}z^i$. Since $-g_0h_0 = 1$, $\tilde{h}(z)$ is monic. Let

$$\mathbf{v} = (0, 0, \cdots, 0, g_{n-k}, \cdots, g_0), \quad \mathbf{u} = (0, 0, \cdots, 0, h_0, \cdots, h_k),$$

and let $\mathbf{v}^l, \mathbf{u}^l$ be the vectors l times cyclicly shifted, that is

$$\mathbf{u}^l = (h_{k-l+1}, h_{k-l+2}, \cdots, h_k, 0, \cdots, 0, h_0, h_1, \cdots, h_{k-l}),$$

and $\mathbf{v}^l$ similarly. First, we see that

$$\mathbf{v} \cdot \mathbf{u}^l = \sum_{i=0}^{k} g_{k+l-i} h_i = 0$$

for $-k < l < n-k$. Hence,

$$\mathbf{v}^m \cdot \mathbf{u}^l = \mathbf{v} \cdot \mathbf{u}^{l-m} = 0$$

for $0 \le m < k$ and $0 \le l < n-k$; that is, each basis vector for C is orthogonal to each basis vector in the code $\tilde{C}$ generated by $\tilde{h}(z)$, and so $\tilde{C} = C^{\perp}$. □

The polynomial $g(z)$ of degree m is called *primitive* if the least positive n such that $g(z)$ divides $z^n - 1$ is $n = (q^m - 1)/(q-1)$. The cyclic code C generated by a primitive $g(z)$ of degree m is a $\left[\frac{q^m-1}{q-1}, \frac{q^m-1}{q-1} - m; q\right]$ *Hamming code.*

The code obtained by shortening the cyclic $[n, k; q]$ code C m times is the $[n-m, k-m; q]$ code

$$\{v(z)g(z) \mid \deg(v(z)) < k'\} \tag{1.1}$$

where $k' = k - m$. Note that (1.1) defines an $[n-k+k', k'; q]$ code for all $k' > 0$, not only for $k' \le k$. These codes are also known as *cyclic redundancy-check* (CRC) codes. The dual of an $[n-k+k', k'; q]$ code C generated by $g(z) = \sum_{i=0}^{n-k} g_i z^i$ where $g_{n-k} = 1$ can be described as a systematic code as follows: The information sequence $(a_{n-k-1}, \ldots, a_0)$ is encoded into the sequence $(a_{n-k-1+k'}, \ldots, a_0)$ where

$$a_j = -\sum_{i=0}^{n-k-1} g_i a_{j-n+k+i}.$$

This follows from the fact that

$$(a_{n-k-1+k'}, a_{n-k-2+k'}, \ldots, a_0) \cdot (0, 0, \ldots, 0, g_{n-k}, \ldots, g_0, \overbrace{0, \ldots, 0}^{i}) = 0$$

for $0 \le i \le k'$ by definition and that $(0, 0, \ldots, 0, g_{n-k}, \ldots, g_0, \overbrace{0, \ldots, 0}^{i})$ where $0 \le i \le k'$ is a basis for C.

A number of binary CRC codes have been selected as international standards for error detection in various contexts. We will return to a more detailed discussion of these and other binary CRC codes in Section 3.5.

1.3 Distance distribution of codes

1.3.1 *Definition of distance distribution*

Let C be an $(n, M; q)$ code. Let

$$A_i = A_i(C) = \frac{1}{M}\#\{(\mathbf{x},\mathbf{y}) \mid \mathbf{x},\mathbf{y} \in C \text{ and } d_{\mathrm{H}}(\mathbf{x},\mathbf{y}) = i\}.$$

The sequence $A_0, A_1, \cdots, A_n$ is known as the *distance distribution* of C and

$$A_C(z) = \sum_{i=0}^{n} A_i z^i$$

is the *distance distribution function* of C.

We will give a couple of alternative expressions for the distance distribution function that will be useful in the study of the probability of undetected error for error detecting codes.

1.3.2 *The MacWilliams transform*

Let C be an $(n, M; q)$ code. The *MacWilliams transform* of $A_C(z)$ is defined by

$$A_C^{\perp}(z) = \frac{1}{M}(1+(q-1)z)^n A_C\left(\frac{1-z}{1+(q-1)z}\right). \tag{1.2}$$

Clearly, $A_C^{\perp}(z)$ is a polynomial in z and we denote the coefficients of $A_C^{\perp}(z)$ by $A_i^{\perp} = A_i^{\perp}(C)$, that is,

$$A_C^{\perp}(z) = \sum_{i=0}^{n} A_i^{\perp} z^i.$$

In particular, $A_0^{\perp} = 1$.

The reason we use the notation $A_C^{\perp}(z)$ is that if C is a linear code, then $A_C^{\perp}(z) = A_{C^{\perp}}(z)$ as we will show below (Theorem 1.14). However, $A_C^{\perp}(z)$ is sometimes useful even if C is not linear. The least $i > 0$ such that $A_i^{\perp}(C) \neq 0$ is known as the *dual distance* $d^{\perp}(C)$.

Substituting $\frac{1-z}{1+(q-1)z}$ for z in the definition of $A_C^{\perp}(z)$ we get the following inverse relation.

Lemma 1.1.

$$A_C(z) = \frac{M}{q^n}(1+(q-1)z)^n A_C^{\perp}\left(\frac{1-z}{1+(q-1)z}\right). \tag{1.3}$$

Differentiating the polynomial (1.3) s times and putting $z = 1$ we get the following relations which are known as the *Pless identities*.

Theorem 1.3. *Let C be an $(n, M; q)$ code and $s \geq 0$. Then*

$$\sum_{i=0}^{n} A_i \binom{i}{s} = \frac{M}{q^s} \sum_{j=0}^{s} A_j^{\perp} (-1)^j (q-1)^{s-j} \binom{n-j}{s-j}.$$

In particular, if $s < d^{\perp}$, then

$$\sum_{i=0}^{n} A_i \binom{i}{s} = \frac{M(q-1)^s}{q^s} \binom{n}{s}.$$

From (1.2) we similarly get the following relation.

Theorem 1.4. *Let C be an $(n, M; q)$ code and $s \geq 0$. Then*

$$\sum_{i=0}^{n} A_i^{\perp} \binom{i}{s} = \frac{q^{n-s}}{M} \sum_{j=0}^{s} A_j (-1)^j (q-1)^{s-j} \binom{n-j}{s-j}.$$

In particular, if $s < d$, then

$$\sum_{i=0}^{n} A_i^{\perp} \binom{i}{s} = \frac{q^{n-s}(q-1)^s}{M} \binom{n}{s}.$$

Two important relations are the following.

Theorem 1.5. *Let C be an $(n, M; q)$ code over Z_q and let $\zeta = e^{2\pi\sqrt{-1}/q}$. Then*

$$A_i^{\perp}(C) = \frac{1}{M^2} \sum_{\substack{\mathbf{u} \in Z_q^n \\ w_H(\mathbf{u})=i}} \Big| \sum_{\mathbf{c} \in C} \zeta^{\mathbf{u} \cdot \mathbf{c}} \Big|^2$$

for $0 \leq i \leq n$.

Note that $\zeta^q = 1$, but $\zeta^j \neq 1$ for $0 < j < q$. Before we prove Theorem 1.5, we first give two lemmas.

Lemma 1.2. *Let $v \in Z_q$. Then*

$$\sum_{u \in Z_q} \zeta^{uv} x^{w_H(u)} = \begin{cases} 1 + (q-1)x & \textit{if } v = 0, \\ 1 - x & \textit{if } v \neq 0. \end{cases}$$

Proof. We have

$$\sum_{u \in Z_q} \zeta^{uv} x^{w_H(u)} = 1 + x \sum_{u=1}^{q-1} \zeta^{uv}.$$

If $v = 0$, the sum is clearly $1 + x(q-1)$. If $v \neq 0$, then

$$\sum_{u=1}^{q-1} \zeta^{uv} = -1 + \sum_{u=0}^{q-1} (\zeta^v)^u = -1 + \frac{1-\zeta^{vq}}{1-\zeta^v} = -1.$$

□

Lemma 1.3. *Let* $\mathbf{v} \in Z_q$. *Then*

$$\sum_{\mathbf{u} \in Z_q^n} \zeta^{\mathbf{u}\cdot\mathbf{v}} x^{w_H(\mathbf{u})} = (1-x)^{w_H(\mathbf{v})}(1+(q-1)x)^{n-w_H(\mathbf{v})}.$$

Proof. From the previous lemma we get

$$\begin{aligned} &\sum_{\mathbf{u} \in Z_q^n} \zeta^{\mathbf{u}\cdot\mathbf{v}} x^{w_H(\mathbf{u})} \\ &= \sum_{u_1 \in Z_q} \zeta^{u_1 v_1} x^{w_H(u_1)} \sum_{u_2 \in Z_q} \zeta^{u_2 v_2} x^{w_H(u_2)} \cdots \sum_{u_n \in Z_q} \zeta^{u_n v_n} x^{w_H(u)} \\ &= (1-x)^{w_H(\mathbf{v})}(1+(q-1)x)^{n-w_H(\mathbf{v})}. \end{aligned}$$

□

We can now prove Theorem 1.5.

Proof. Since $d_H(\mathbf{c}, \mathbf{c}') = w_H(\mathbf{c} - \mathbf{c}')$, Lemma 1.3 gives

$$\begin{aligned} \sum_{i=0}^{n} A_i^{\perp} x^i &= \frac{1}{M} \sum_{i=0}^{n} A_i (1-x)^i (1+(q-1)x)^{n-i} \\ &= \frac{1}{M^2} \sum_{\mathbf{c} \in C} \sum_{\mathbf{c}' \in C} (1-x)^{d_H(\mathbf{c},\mathbf{c}')} (1+(q-1)x)^{n-d_H(\mathbf{c},\mathbf{c}')} \\ &= \frac{1}{M^2} \sum_{\mathbf{c} \in C} \sum_{\mathbf{c}' \in C} \sum_{\mathbf{u} \in Z_q^n} \zeta^{\mathbf{u}\cdot(\mathbf{c}-\mathbf{c}')} x^{w_H(\mathbf{u})} \\ &= \frac{1}{M^2} \sum_{\mathbf{u} \in Z_q^n} x^{w_H(\mathbf{u})} \sum_{\mathbf{c} \in C} \zeta^{\mathbf{u}\cdot\mathbf{c}} \sum_{\mathbf{c}' \in C} \zeta^{-\mathbf{u}\cdot\mathbf{c}'}. \end{aligned}$$

Observing that

$$\sum_{\mathbf{c} \in C} \zeta^{\mathbf{u}\cdot\mathbf{c}} \sum_{\mathbf{c}' \in C} \zeta^{-\mathbf{u}\cdot\mathbf{c}'} = \Big| \sum_{\mathbf{c} \in C} \zeta^{\mathbf{u}\cdot\mathbf{c}} \Big|^2,$$

the theorem follows. □

When the alphabet is $GF(q)$, there is a similar expression for $A_i^{\perp}(C)$. Let $q = p^r$, where p is a prime. The *trace function* from $GF(q)$ to $GF(p)$ is defined by

$$Tr(a) = \sum_{i=0}^{r-1} a^{p^i}.$$

One can show that $Tr(a) \in GF(p)$ for all $a \in GF(q)$, and that $Tr(a+b) = Tr(a) + Tr(b)$.

Theorem 1.6. *Let $q = p^r$ where p is a prime. Let C be an $(n, M; q)$ code over $GF(q)$ and let $\zeta = e^{2\pi\sqrt{-1}/p}$. Then*

$$A_i^{\perp}(C) = \frac{1}{M^2} \sum_{\substack{\mathbf{u} \in GF(q)^n \\ w_H(\mathbf{u})=i}} \left| \sum_{\mathbf{c} \in C} \zeta^{Tr(\mathbf{u} \cdot \mathbf{c})} \right|^2$$

for $0 \le i \le n$.

The proof is similar to the proof of Theorem 1.5.

Corollary 1.1. *Let C be an $(n, M; q)$ code (over Z_q or over $GF(q)$). Then*

$$A_i^{\perp}(C) \ge 0 \text{ for } 0 \le i \le n.$$

1.3.3 *Binomial moment*

We have

$$\begin{aligned}
\sum_{j=1}^{n} A_j x^j &= \sum_{j=1}^{n} A_j x^j (x + 1 - x)^{n-j} \\
&= \sum_{j=1}^{n} A_j x^j \sum_{l=0}^{n-j} \binom{n-j}{l} x^l (1-x)^{n-j-l} \\
&= \sum_{i=1}^{n} x^i (1-x)^{n-i} \sum_{j=1}^{i} A_j \binom{n-j}{i-j}. \qquad (1.4)
\end{aligned}$$

The *binomial moment* is defined by

$$A_i^{\diamond}(C) = \sum_{j=1}^{i} A_j(C) \binom{n-j}{n-i}$$

for $1 \le i \le n$.

The relation (1.4) then can be expressed as follows:

Theorem 1.7. *Let C be an $(n, M; q)$ code. Then*

$$A_C(x) = 1 + \sum_{i=1}^{n} A_i^{\diamond}(C)x^i(1-x)^{n-i}.$$

We note that the A_i can be expressed in terms of the $A_j^{\diamond}$.

Theorem 1.8.

$$A_i(C) = \sum_{j=1}^{i}(-1)^{j-i}A_j^{\diamond}(C)\binom{n-j}{n-i}$$

for $1 \le i \le n$.

Proof.

$$\begin{aligned}
\sum_{j=1}^{i}(-1)^{j-i}A_j^{\diamond}(C)\binom{n-j}{n-i} &= \sum_{j=1}^{i}(-1)^{j-i}\binom{n-j}{n-i}\sum_{k=1}^{j}A_k(C)\binom{n-k}{n-j} \\
&= \sum_{k=1}^{i}A_k(C)\sum_{j=k}^{i}(-1)^{j-i}\binom{n-j}{n-i}\binom{n-k}{n-j} \\
&= \sum_{k=1}^{i}A_k(C)\sum_{j=k}^{i}(-1)^{j-i}\binom{n-k}{n-i}\binom{i-k}{i-j} \\
&= \sum_{k=1}^{i}A_k(C)\binom{n-k}{n-i}(-1)^{i-k}\sum_{j=k}^{i}(-1)^{j-k}\binom{i-k}{j-k} \\
&= \sum_{k=1}^{i}A_k(C)\binom{n-k}{n-i}(-1+1)^{i-k} \\
&= A_i(C).
\end{aligned}$$

□

We can also express $A_i^\diamond$ in terms of the $A_j^\perp$. We have

$$
\begin{aligned}
A_C(x)-1 &= \frac{M}{q^n}\sum_{j=0}^{n} A_j^\perp (1-x)^j (1+(q-1)x)^{n-j} - 1\\
&= \frac{M}{q^n}\sum_{j=0}^{n} A_j^\perp (1-x)^j (qx+1-x)^{n-j} - (x+1-x)^n\\
&= \frac{M}{q^n}\sum_{j=0}^{n} A_j^\perp (1-x)^j \sum_{i=0}^{n-j}\binom{n-j}{i} q^i x^i (1-x)^{n-j-i}\\
&\quad -\sum_{i=0}^{n}\binom{n}{i} x^i (1-x)^{n-i}\\
&= \sum_{i=0}^{n} x^i (1-x)^{n-i}\left\{\frac{M}{q^n} q^i \sum_{j=0}^{n-i}\binom{n-j}{i} A_j^\perp - \binom{n}{i}\right\}.
\end{aligned}
$$

Hence we get the following result.

Theorem 1.9. *Let C be an $(n, M; q)$ code. Then, for $1 \le i \le n$,*

$$
\begin{aligned}
A_i^\diamond(C) &= Mq^{i-n}\sum_{j=0}^{n-i}\binom{n-j}{i} A_j^\perp - \binom{n}{i}\\
&= \binom{n}{i}(Mq^{i-n}-1) + Mq^{i-n}\sum_{j=d^\perp}^{n-i}\binom{n-j}{i} A_j^\perp .
\end{aligned}
$$

From the definition and Theorem 1.9 we get the following corollary.

Corollary 1.2. *Let C be an $(n, M; q)$ code with minimum distance d and dual distance $d^\perp$. Then*

$$A_i^\diamond(C) = 0 \text{ for } 1 \le i \le d-1,$$

$$A_i^\diamond(C) \ge \max\left\{0, \binom{n}{i}(Mq^{i-n}-1)\right\} \text{ for } d \le i \le n-d^\perp,$$

and

$$A_i^\diamond(C) = \binom{n}{i}(Mq^{i-n}-1) \text{ for } n-d^\perp < i \le n.$$

There is an alternative expression for $A_i^\diamond(C)$ which is more complicated, but quite useful.

For each set $E \subset \{1, 2, \ldots, n\}$, define an equivalence relation $\sim_E$ on C by $\mathbf{x} \sim_E \mathbf{y}$ if and only if $\chi(\mathbf{x}, \mathbf{y}) \subseteq E$ (that is, $x_i = y_i$ for all $i \notin E$). Let the set of equivalence classes be denoted X_E. If two vectors differ in at least one position outside E, then they are not equivalent. Therefore, the number of equivalence classes, that is, the size of X_E, is $q^{n-\#E}$.

Theorem 1.10. *Let C be an $(n, M; q)$ code. Then, for $1 \le i \le n$,*

$$A_j^\diamond(C) = \frac{1}{M} \sum_{\substack{E \subset \{1,2,\ldots,n\} \\ \#E = j}} \sum_{U \in X_E} \#U(\#U - 1).$$

Proof. We count the number of elements in the set

$$V = \{(E, \mathbf{x}, \mathbf{y}) \mid E \subset \{1, 2, \ldots, n\}, \#E = j, \mathbf{x}, \mathbf{y} \in C, \mathbf{x} \ne \mathbf{y}, \mathbf{x} \sim_E \mathbf{y}\}$$

in two ways. On one hand, for given E and an equivalence class $U \in X_E$, the pair $(\mathbf{x}, \mathbf{y})$ can be chosen in $\#U(\#U - 1)$ different ways. Hence, the the number of elements of V is given by

$$\#V = \sum_{\substack{E \subset \{1,2,\ldots,n\} \\ \#E = j}} \sum_{U \in X_E} \#U(\#U - 1). \tag{1.5}$$

On the other hand, for a given pair $(\mathbf{x}, \mathbf{y})$ of code words at distance $i \le j$, E must contain the i elements in the support $\chi(\mathbf{x}, \mathbf{y})$ and $j - i$ of the $n - i$ elements outside the support. Hence, E can be chosen in $\binom{n-i}{j-i}$ ways. Since a pair $(\mathbf{x}, \mathbf{y})$ of code words at distance i can be chosen in $MA_i(C)$ ways, we get

$$\#V = \sum_{i=1}^{j} MA_i(C)\binom{n-i}{j-i} = MA_j^\diamond(C). \tag{1.6}$$

Theorem 1.10 follows by combining (1.5) and (1.6). □

From Theorem 1.10 we can derive a lower bound on $A_j^\diamond(C)$ which is sharper than (or sometimes equal to) the bound in Corollary 1.2.

First we need a simple lemma.

Lemma 1.4. *Let $m_1, m_2, \ldots, m_N$ be non-negative integers with sum M. Then*

$$\sum_{i=1}^{N} m_i^2 \ge \left(2\left\lfloor\frac{M}{N}\right\rfloor + 1\right)\left(M - N\left\lfloor\frac{M}{N}\right\rfloor\right) + N\left\lfloor\frac{M}{N}\right\rfloor^2 \tag{1.7}$$

$$= M + \left(\left\lceil\frac{M}{N}\right\rceil - 1\right)\left(2M - N\left\lceil\frac{M}{N}\right\rceil\right), \tag{1.8}$$

with equality if and only if

$$\left\lfloor\frac{M}{N}\right\rfloor \le m_i \le \left\lceil\frac{M}{N}\right\rceil$$

for all i.

Proof. Let $x_1, x_2, \ldots, x_N$ be non-negative integers for which $\sum_{i=1}^{N} x_i^2$ is minimal. Without loss of generality, we may assume that $x_1 \le x_i \le x_N$ for all i. Suppose $x_N \ge x_1 + 2$. Let $y_1 = x_1 + 1$, $y_N = x_N - 1$, $y_i = x_i$ otherwise. Then, by the minimality of $\sum x_i^2$,

$$0 \le \sum_{i=1}^{N} y_i^2 - \sum_{i=1}^{N} x_i^2 = (x_1+1)^2 - x_1^2 + (x_N - 1)^2 - x_N^2 = 2(x_1 - x_N + 1),$$

contradicting the assumption $x_N \ge x_1 + 2$. Therefore, we must have

$$x_N = x_1 + 1 \text{ or } x_N = x_1.$$

Let $\alpha = \lfloor M/N \rfloor$ and $M = N\alpha + \beta$ where $0 \le \beta < N$. Then β of the x_i must have value $\alpha + 1$ and the remaining $N - \beta$ have value α and so

$$\sum_{i=1}^{N} x_i^2 = \beta(\alpha+1)^2 + (N-\beta)\alpha^2 = (2\alpha+1)\beta + N\alpha^2.$$

This proves (1.7). We have

$$\left\lceil \frac{M}{N} \right\rceil = \alpha \text{ if } \beta = 0, \text{ and } \left\lceil \frac{M}{N} \right\rceil = \alpha + 1 \text{ if } \beta > 0.$$

Hence (1.8) follows by rewriting (1.7). □

Using Lemma 1.4, with the lower bound in the version (1.8), we see that the inner sum $\sum_{U \in X_E} \#U(\#U - 1)$ in Theorem 1.7 is lower bounded by

$$\sum_{U \in X_E} \#U(\#U-1) \ge \left(\left\lceil \frac{M}{q^{n-j}} \right\rceil - 1 \right) \left(2M - q^{n-j} \left\lceil \frac{M}{q^{n-j}} \right\rceil \right),$$

independent of E. For E there are $\binom{n}{j}$ possible choices. Hence, we get the following bound.

Theorem 1.11. *Let C be an $(n, M; q)$ code. Then, for $1 \le j \le n$,*

$$A_j^{\circ}(C) \ge \binom{n}{j} \left(\left\lceil \frac{M}{q^{n-j}} \right\rceil - 1 \right) \left(2 - \frac{q^{n-j}}{M} \left\lceil \frac{M}{q^{n-j}} \right\rceil \right).$$

1.3.4 *Distance distribution of complementary codes*

There is a close connection between the distance distributions of a code and its (set) complement. More general, there is a connection between the distance distributions of two disjoint codes whose union is a distance invariant code.

An $(n, M; q)$ code is called *distance invariant* if

$$\sum_{\mathbf{y}\in C} z^{d_H(\mathbf{x},\mathbf{y})} = A_C(z)$$

for all $\mathbf{x} \in C$. In particular, any linear code is distance invariant. However, a code may be distance invariant without being linear.

Example 1.1. A simple example of a non-linear distance invariant code is the code

$$\{(1000), (0100), (0010), (0001)\}.$$

Theorem 1.12. *Let the $(n, M_1; q)$ code C_1 and the $(n, M_2; q)$ code C_2 be disjoint codes such that $C_1 \cup C_2$ is distance invariant. Then,*

$$M_1\Big\{A_{C_1\cup C_2}(z) - A_{C_1}(z)\Big\} = M_2\Big\{A_{C_1\cup C_2}(z) - A_{C_2}(z)\Big\}.$$

Proof. Since $C_1 \cup C_2$ is distance invariant, we have

$$\begin{aligned} M_1 A_{C_1\cup C_2}(z) &= \sum_{\mathbf{x}\in C_1}\sum_{\mathbf{y}\in C_1\cup C_2} z^{d_H(\mathbf{x},\mathbf{y})} \\ &= \sum_{\mathbf{x}\in C_1}\sum_{\mathbf{y}\in C_1} z^{d_H(\mathbf{x},\mathbf{y})} + \sum_{\mathbf{x}\in C_1}\sum_{\mathbf{y}\in C_2} z^{d_H(\mathbf{x},\mathbf{y})} \\ &= M_1 A_{C_1}(z) + \sum_{\mathbf{x}\in C_1}\sum_{\mathbf{y}\in C_2} z^{d_H(\mathbf{x},\mathbf{y})}, \end{aligned}$$

and so

$$M_1\Big\{A_{C_1\cup C_2}(z) - A_{C_1}(z)\Big\} = \sum_{\mathbf{x}\in C_1}\sum_{\mathbf{y}\in C_2} z^{d_H(\mathbf{x},\mathbf{y})}.$$

Similarly,

$$M_2\Big\{A_{C_1\cup C_2}(z) - A_{C_2}(z)\Big\} = \sum_{\mathbf{x}\in C_1}\sum_{\mathbf{y}\in C_2} z^{d_H(\mathbf{x},\mathbf{y})},$$

and the theorem follows. □

If $C_2 = \overline{C_1}$, then the conditions of Theorem 1.12 are satisfied. Since $C_1 \cup C_2 = F_q^n$ we have $M_2 = q^n - M_1$ and $A_{C_1\cup C_2}(z) = (1 + (q-1)z)^n$. Hence we get the following corollary.

Corollary 1.3. *Let C be an $(n, M; q)$ code. Then*

$$A_{\overline{C}}(z) = \frac{M}{q^n - M} A_C(z) + \frac{q^n - 2M}{q^n - M}(1 + (q-1)z)^n.$$

From Corollary 1.3 we immediately get the following corollary.

Corollary 1.4. *Let C be an $(n, M; q)$ code. Then, for $0 \le i \le n$, we have*

$$A_i(\overline{C}) = \frac{M}{q^n - M} A_i(C) + \frac{q^n - 2M}{q^n - M} \binom{n}{i} (q-1)^i.$$

Using Corollary 1.4 we get the following.

Corollary 1.5. *Let C be an $(n, M; q)$ code. Then, for $1 \le i \le n$, we have*

$$A_i^{\diamond}(\overline{C}) = \frac{M}{q^n - M} A_i^{\diamond}(C) + \frac{q^n - 2M}{q^n - M} \binom{n}{i} (q^i - 1).$$

Proof. We have

$$\begin{aligned} A_i^{\diamond}(\overline{C}) &= \sum_{j=1}^{i} A_i(\overline{C}) \binom{n-j}{n-i} \\ &= \frac{M}{q^n - M} \sum_{j=1}^{i} A_j(C) \binom{n-j}{n-i} + \frac{q^n - 2M}{q^n - M} \sum_{j=1}^{i} \binom{n-j}{n-i} \binom{n}{j} (q-1)^j \\ &= \frac{M}{q^n - M} A_i^{\diamond}(C) + \frac{q^n - 2M}{q^n - M} \sum_{j=1}^{i} \binom{n}{i} \binom{i}{j} (q-1)^j \\ &= \frac{M}{q^n - M} A_i^{\diamond}(C) + \frac{q^n - 2M}{q^n - M} \binom{n}{i} (q^i - 1). \end{aligned}$$

□

1.4 Weight distribution of linear codes

1.4.1 *Weight distribution*

Let

$$A_i^{\mathrm{w}} = A_i^{\mathrm{w}}(C) = \#\{\mathbf{x} \in C \mid w_{\mathrm{H}}(\mathbf{x}) = i\}.$$

The sequence $A_0^{\mathrm{w}}, A_1^{\mathrm{w}}, \cdots, A_n^{\mathrm{w}}$ is known as the *weight distribution* of C and

$$A_C^{\mathrm{w}}(z) = \sum_{i=0}^{n} A_i^{\mathrm{w}} z^i$$

is the *weight distribution function* of C.

We note that $d_{\mathrm{H}}(\mathbf{x}, \mathbf{y}) = w_{\mathrm{H}}(\mathbf{x} - \mathbf{y})$. If C is linear, then $\mathbf{x} - \mathbf{y} \in C$ when $\mathbf{x}, \mathbf{y} \in C$. Hence we get the following useful result.

Theorem 1.13. *For a linear code C we have $A_i(C) = A_i^{\mathrm{w}}(C)$ for all i and $A_C(z) = A_C^{\mathrm{w}}(z)$.*

If C and C' are equivalent codes, then clearly $A_i(C) = A_i(C')$. In particular, for the study of the weight distribution of linear codes we may therefore without loss of generality assume that the code is systematic if we so wish.

1.4.2 *Weight distribution of $*$-extended codes*

The $*$-operation for linear codes was defined on page 8. The code S_k is a constant weight code, that is, all non-zero code words have the same weight, namely q^{k-1}.

Therefore, $A_{C^*}(z)$ only depends on $A_C(z)$. In fact

$$A_{C^*}(z) - 1 = z^{q^{k-1}}(A_C(z) - 1)$$

since each non-zero vector is extended by a part of weight q^{k-1}.

1.4.3 *MacWilliams's theorem*

The following theorem is known as *MacWilliams's theorem.*

Theorem 1.14. *Let C be a linear $[n, k; q]$ code. Then*

$$A_i^{\perp}(C) = A_i(C^{\perp}).$$

Equivalently,

$$A_{C^{\perp}}(z) = \frac{1}{q^k}(1 + (q-1)z)^n A_C\left(\frac{1-z}{1+(q-1)z}\right).$$

Proof. We prove this for q a prime, using Theorem 1.5. The proof for general prime power q is similar, using Theorem 1.6. First we show that

$$\sum_{\mathbf{c}\in C} \zeta^{\mathbf{u}\cdot\mathbf{c}} = \begin{cases} M \text{ if } \mathbf{u} \in C^{\perp}, \\ 0 \;\text{ if } \mathbf{u} \notin C^{\perp}. \end{cases} \tag{1.9}$$

If $\mathbf{u} \in C^{\perp}$, then $\mathbf{u}\cdot\mathbf{c} = 0$ and $\zeta^{\mathbf{u}\cdot\mathbf{c}} = 1$ for all $\mathbf{c} \in C$, and the result follows. If $\mathbf{u} \notin C^{\perp}$, then there exists a code word $\mathbf{c}' \in C$ such that $\mathbf{u}\cdot\mathbf{c}' \neq 0$ and hence $\zeta^{\mathbf{u}\cdot\mathbf{c}'} \neq 1$. Because of the linearity, $\mathbf{c} + \mathbf{c}'$ runs through C when $\mathbf{c}$ does. Hence

$$\sum_{\mathbf{c}\in C} \zeta^{\mathbf{u}\cdot\mathbf{c}} = \sum_{\mathbf{c}\in C} \zeta^{\mathbf{u}\cdot(\mathbf{c}+\mathbf{c}')} = \zeta^{\mathbf{u}\cdot\mathbf{c}'} \sum_{\mathbf{c}\in C} \zeta^{\mathbf{u}\cdot\mathbf{c}}.$$

Hence $\sum_{\mathbf{c}\in C} \zeta^{\mathbf{u}\cdot\mathbf{c}} = 0$. This proves (1.9). By Theorem 1.5,

$$A_i^{\perp}(C) = \frac{1}{M^2} \sum_{\substack{\mathbf{u}\in Z_q^n \\ w_H(\mathbf{u})=i}} \Big|\sum_{\mathbf{c}\in C} \zeta^{\mathbf{u}\cdot\mathbf{c}}\Big|^2 = \frac{1}{M^2} \sum_{\substack{\mathbf{u}\in C^{\perp} \\ w_H(\mathbf{u})=i}} M^2 = A_i(C^{\perp}).$$

□

Corollary 1.6. *Let C be a linear $[n, k; q]$ code. Then*

$$d^{\perp}(C) = d(C^{\perp}).$$

1.4.4 *A generalized weight distribution*

Many generalizations of the weight distribution have been studied. One that is particularly important for error detection is the following.

Let C be an $[n, k]$ code and m a divisor of n. Let $A_{i_1,i_2,\cdots,i_m}(C)$ be the number of vectors $(\mathbf{x}_1|\mathbf{x}_2|\cdots|\mathbf{x}_m) \in C$ such that each part $\mathbf{x}_j \in GF(q)^{n/m}$ and $w_{\mathrm{H}}(\mathbf{x}_j) = i_j$ for $j = 1, 2, \cdots, m$. Further, let

$$A_C(z_1, z_2, \cdots, z_m) = \sum_{i_1,i_2,\cdots,i_m} A_{i_1,i_2,\cdots,i_m}(C) z_1^{i_1} z_2^{i_2} \cdots z_m^{i_m}.$$

For $m = 1$ we get the usual weight distribution function. Theorem 1.14 generalizes as follows.

Theorem 1.15. *Let C be a linear $[n, k; q]$ code. Then*

$$A_C(z_1, z_2, \cdots, z_m) = q^{k-n} \left\{ \prod_{j=1}^{m} (1 + (q-1)z_j) \right\}^{\frac{n}{m}} A_{C^{\perp}}(z_1', z_2', \cdots z_m')$$

where

$$z_j' = \frac{1 - z_j}{1 + (q-1)z_j}.$$

1.4.5 *Linear codes over larger fields*

There is an alternative expression for the weight distribution function that is useful for some applications. Let G be a $k \times n$ generator matrix over $GF(q)$. Let $m_G : GF(q)^k \to \mathcal{N} = \{0, 1, 2, \ldots\}$, the *column count function*, be defined such that G contains exactly $m(\mathbf{x}) = m_G(\mathbf{x})$ columns equal to $\mathbf{x}$ for all $\mathbf{x} \in GF(q)^k$. We use the following further notations:

$[a]_b = \prod_{i=0}^{b-1}(q^a - q^i)$,

$s(U, m) = \sum_{\mathbf{x} \in U} m(\mathbf{x})$ for all $U \subseteq GF(q)^k$,

$\mathcal{S}_{kl}$ is the set of l dimensional subspaces of $GF(q)^k$,

$\sigma_{kl}(m, z) = \sum_{U \in \mathcal{S}_{kl}} z^{s(\bar{U}, m)}$, where $\bar{U} = GF(q)^k \setminus U$,

$\hat{U} = \{\mathbf{y} \in GF(q^r)^k \mid \mathbf{y} \cdot \mathbf{x} = 0 \text{ for } \mathbf{x} \in GF(q)^k \text{ if and only if } \mathbf{x} \in U\}$,

$C_r = \{\mathbf{y}G \mid \mathbf{y} \in GF(q^r)^k\}$, the code generated by G over $GF(q^r)$,

$C = C_1$.

Theorem 1.16. *For $r \geq 1$ we have*

$$A_{C_r}(z) = \sum_{l=0}^{k} [r]_{k-l} \sigma_{kl}(m, z).$$

Proof. First we note that if $\mathbf{y} \in \hat{U}$, then

$$w_{\mathrm{H}}(\mathbf{y}G) = \sum_{\mathbf{x} \in GF(q)^k} m(\mathbf{x}) w_{\mathrm{H}}(\mathbf{y} \cdot \mathbf{x}) = \sum_{\mathbf{x} \in \bar{U}} m(\mathbf{x}) = s(\bar{U}, m).$$

Hence

$$A_{C_r}(z) = \sum_{l=0}^{k} \sum_{U \in \mathcal{S}_{kl}} \sum_{\mathbf{y} \in \hat{U}} z^{w_{\mathrm{H}}(\mathbf{y}G)} = \sum_{l=0}^{k} \sum_{U \in \mathcal{S}_{kl}} z^{s(\bar{U}, m)} \sum_{\mathbf{y} \in \hat{U}} 1.$$

Since

$$\sum_{\mathbf{y} \in \hat{U}} 1 = [r]_{k-l},$$

the theorem follows. □

For $r = 1$, we get the following alternative expression for the weight distribution of C.

Corollary 1.7. *We have*

$$A_C(z) = 1 + \sum_{U \in \mathcal{S}_{k,k-1}} z^{s(\bar{U}, m)}.$$

1.4.6 *Weight distribution of cosets*

Theorem 1.17. *Let C be an $[n, k; q]$ code and S a proper coset of C. Let D be the $[n, k+1; q]$ code containing C and S. Then*

$$A_S^{\mathrm{w}}(z) = \frac{1}{q-1} \Big\{ A_D(z) - A_C(z) \Big\}.$$

Proof. For each non-zero $a \in GF(q)$, $aS = \{a\mathbf{x} \mid \mathbf{x} \in S\}$ is also a proper coset of C and $A^{\mathrm{w}}_{aS}(z) = A^{\mathrm{w}}_{S}(z)$. Further $D = C \cup \bigcup_{a \neq 0} aS$ (disjoint union) and so

$$A_D(z) = A_C(z) + (q-1)A^{\mathrm{w}}_S(z), \tag{1.10}$$

and the theorem follows. □

Using the MacWilliams identity we get the following alternative expression.

Corollary 1.8. *Let C be an $[n, k; q]$ code and S a proper coset of C. Let D be the $[n, k+1; q]$ code containing C and S. Then*

$$A^{\mathrm{w}}_S(z) = \frac{\Big(1+(q-1)z\Big)^n}{q^{n-k}(q-1)} \left\{ qA_{D^\perp}\left(\frac{1-z}{1+(q-1)z}\right) - A_{C^\perp}\left(\frac{1-z}{1+(q-1)z}\right)\right\}.$$

Theorem 1.18. *Let C be an $[n, k; q]$ code and S a proper coset of C. Then*

$$A^{\mathrm{w}}_S(z) \geq z^{n-k} A_C(z) \tag{1.11}$$

for all $z \in [0, 1]$.

Proof. We may assume without loss of generality that the code C is systematic. There exists a $\mathbf{v} \in S$ such that $S = \mathbf{v} + C$ and such that $\mathbf{v} = (\mathbf{0}|\mathbf{b})$ where $\mathbf{b} \in GF(q)^{n-k}$.

Let $(\mathbf{x}|\mathbf{x}') \in C$ where $\mathbf{x} \in GF(q)^k$ and $\mathbf{x}' \in GF(q)^{n-k}$. Then

$$\begin{aligned} w_{\mathrm{H}}((\mathbf{x}|\mathbf{x}') + (\mathbf{0}|\mathbf{b})) &= w_{\mathrm{H}}(\mathbf{x}) + w_{\mathrm{H}}(\mathbf{x}' + \mathbf{b}) \\ &\leq w_{\mathrm{H}}(\mathbf{x}) + n - k \\ &\leq w_{\mathrm{H}}((\mathbf{x}|\mathbf{x}')) + n - k \end{aligned}$$

and so

$$z^{w_{\mathrm{H}}((\mathbf{x}|\mathbf{x}')+(\mathbf{0}|\mathbf{b}))} \geq z^{n-k} z^{w_{\mathrm{H}}((\mathbf{x}|\mathbf{x}'))}.$$

Summing over all $(\mathbf{x}|\mathbf{x}') \in C$, the theorem follows. □

Corollary 1.9. *Let C be an $[n, k; q]$ code and D an $[n, k+1; q]$ code containing C. Then*

$$A_D(z) \geq \left\{1 + (q-1)z^{n-k}\right\} A_C(z).$$

Proof. Let $S \subset D$ be a proper coset of C. By (1.10) and Theorem 1.18 we have

$$A_D(z) = A_C(z) + (q-1)A^{\mathrm{w}}_S(z) \geq A_C(z) + (q-1)z^{n-k}A_C(z).$$ □

Theorem 1.19. *Let C be an $[n,k;q]$ code and S a proper coset of C. Then*

$$A_S^{\mathrm{w}}(z) \le \frac{1-y^{k+1}}{1+(q-1)y^{k+1}} A_C(z)$$

for all $z \in [0,1]$, where $y = (1-z)/(1+(q-1)z)$.

Theorem 1.20. *Let C be an $[n,k;q]$ code and D an $[n,k+1;q]$ code which contains C. Then*

$$A_C(z) \ge \frac{1+(q-1)y^{k+1}}{q} A_D(z)$$

for all $z \in [0,1]$, where $y = (1-z)/(1+(q-1)z)$.

Proof. By Corollary 1.9 we get

$$\begin{aligned}
A_C(z) &= q^{k-n}\Big(1+(q-1)z\Big)^n A_{C^\perp}(y) \\
&\ge q^{k-n}\Big(1+(q-1)z\Big)^n \Big(1+(q-1)y^{k+1}\Big) A_{D^\perp}(y) \\
&= q^{-1}\Big(1+(q-1)y^{k+1}\Big) A_D(z) \\
&= q^{-1}\Big(1+(q-1)y^{k+1}\Big)\Big(A_C(z)+(q-1)A_S^{\mathrm{w}}(z)\Big)
\end{aligned}$$

and the theorems follow. □

Corollary 1.10. *If C is an $[n,k;q]$ code and $k < n$, then*

$$A_C(z) \ge \frac{(1+(q-1)z)^n}{q^{n-k}} \prod_{j=k+1}^{n} \left(1+(q-1)y^j\right),$$

for all $z \in [0,1]$, where $y = (1-z)/(1+(q-1)z)$.

Proof. The corollary follows from Theorem 1.20 by induction on k. □

1.4.7 *Counting vectors in a sphere*

The *sphere* $S_t(\mathbf{x})$ of radius t around a vector $\mathbf{x} \in GF(q)^n$ is the set of vectors within Hamming distance t of $\mathbf{x}$, that is

$$S_t(\mathbf{x}) = \{\mathbf{y} \in GF(q)^n \mid d_{\mathrm{H}}(\mathbf{x},\mathbf{y}) \le t\}.$$

Let $N_t(i,j)$ be the number of vectors of weight j in a sphere of radius t around a vector of weight i.

Theorem 1.21. *We have*

$$N_t(i,j) = \sum_{e=|i-j|}^{t} \sum_{\delta=\max(i,j)-e}^{\min(\lfloor \frac{i+j-e}{2} \rfloor, n-e)} \binom{n-i}{\beta} \frac{i!}{\gamma!\delta!\epsilon!}(q-1)^{\beta}(q-2)^{\epsilon}$$

where $\beta = e - i + \delta$, $\gamma = e - j + \delta$, $\epsilon = i + j - e - 2\delta$.

Proof. Let $w_{\mathrm{H}}(\mathbf{x}) = i$ and let $\mathbf{y} \in S_t(\mathbf{x})$ such that $w_{\mathrm{H}}(\mathbf{y}) = j$. Let

$$\begin{aligned} \alpha &= \#\{l \mid x_l = y_l = 0\}, \\ \beta &= \#\{l \mid x_l = 0, y_l \neq 0\}, \\ \gamma &= \#\{l \mid x_l \neq 0, y_l = 0\}, \\ \delta &= \#\{l \mid x_l = y_l \neq 0\}, \\ \epsilon &= \#\{l \mid x_l \neq 0, y_l \neq 0, x_l \neq y_l\}. \end{aligned} \tag{1.12}$$

Then

$$\begin{aligned} i &= w_{\mathrm{H}}(\mathbf{x}) = \gamma + \delta + \epsilon, \\ j &= w_{\mathrm{H}}(\mathbf{y}) = \beta + \delta + \epsilon, \\ e &= d_{\mathrm{H}}(\mathbf{x}, \mathbf{y}) = \beta + \gamma + \epsilon, \\ n &= \alpha + \beta + \gamma + \delta + \epsilon. \end{aligned} \tag{1.13}$$

Hence

$$\begin{aligned} \beta &= e - i + \delta, \\ \gamma &= e - j + \delta, \\ \epsilon &= i + j - e - 2\delta. \end{aligned} \tag{1.14}$$

Further,

$$\begin{aligned} &|i-j| \leq e \leq t, \\ &\delta = i - e + \beta \geq i - e, \\ &\delta = j - e + \gamma \geq j - e, \\ &\delta = n - e - \alpha \leq n - e, \\ &2\delta = i + j - e - \epsilon \leq i + j - e. \end{aligned} \tag{1.15}$$

On the other hand, if e and δ are integers such that (1.15) is satisfied, then there are

$$\binom{n-i}{\beta} \frac{i!}{\gamma!\delta!\epsilon!}(q-1)^{\beta}(q-2)^{\epsilon}$$

ways to choose $\mathbf{y}$ such that (1.12)–(1.14) are satisfied. □

For $q = 2$, the terms in the sum for $N_t(i,j)$ are 0 unless $\epsilon = 0$. We get the following simpler expression in this case:

$$N_t(i,j) = \sum_{\gamma=\max(0,i-j)}^{\lfloor \frac{t+i-j}{2} \rfloor} \binom{n-i}{\gamma+j-i} \binom{i}{\gamma}. \tag{1.16}$$

1.4.8 *Bounds on the number of code words of a given weight*

Some useful upper bounds on A_i for a linear code are given by the next theorem.

Theorem 1.22. *Let C be a linear $[n, k, d = 2t+1; q]$ code. If $N_t(i, j) > 0$, then*

$$A_i \le \frac{\binom{n}{j}}{N_t(i,j)}(q-1)^j.$$

In particular, for $d \le i \le \lfloor \frac{n}{2} \rfloor$ we have

$$A_i \le \frac{\binom{n}{i}}{\binom{n-i+t}{t}}(q-1)^{i-t} \le \frac{\binom{n}{i}}{\binom{\lceil \frac{n}{2} \rceil + t}{t}}(q-1)^{i-t},$$

and, for $\lceil \frac{n}{2} \rceil \le i \le n-t$,

$$A_i \le \frac{\binom{n}{i}}{\binom{i+t}{t}}(q-1)^i \le \frac{\binom{n}{i}}{\binom{\lceil \frac{n}{2} \rceil + t}{t}}(q-1)^i.$$

Proof. Counting all vectors of weight j and Hamming distance at most t from a code word of weight i we get

$$A_i N_t(i,j) \le \binom{n}{j}(q-1)^j.$$

In particular, $N_t(i, i-t) = \binom{i}{t} > 0$ for all $i \ge d$, and so

$$A_i \le \frac{\binom{n}{i-t}}{\binom{i}{t}}(q-1)^{i-t} = \frac{\binom{n}{i}}{\binom{n-i+t}{t}}(q-1)^{i-t}.$$

Similarly, $N_t(i, i+t) = \binom{n-i}{t}(q-1)^t > 0$ for $d \le i \le n-t$ and so

$$A_i \le \frac{\binom{n}{i+t}(q-1)^{i+t}}{\binom{n-i}{t}(q-1)^t} = \frac{\binom{n}{i}}{\binom{i+t}{t}}(q-1)^i.$$

□

Theorem 1.23. *For an $[n, k; q]$ code C we have $A_n \le (q-1)^k$.*

Proof. Since equivalent codes have the same weight distribution, we may assume without loss of generality that the code is systematic, that is, it is generated by a matrix

$$G = (I_k | P) = \begin{pmatrix} \mathbf{g}_1 \\ \vdots \\ \mathbf{g}_k \end{pmatrix}$$

where I_k is the $k \times k$ identity matrix, P is a $k \times (n-k)$ matrix, and $\mathbf{g}_1, \ldots, \mathbf{g}_k$ are the rows of G. If $\mathbf{c} = \sum_{i=1}^{k} a_i \mathbf{g}_i$ has weight n, then in particular $a_i = c_i \ne 0$ for $1 \le i \le k$. Hence there are at most $(q-1)^k$ such $\mathbf{c}$. □

There are many codes for which we have $A_n = (q-1)^k$. For example, this is the case for any code that has a generator matrix where all the columns have weight one.

1.5 The weight hierarchy

For a linear $[n,k;q]$ code C and any r, where $1 \le r \le k$, *the r-th minimum support weight* is defined by

$$d_r = d_r(C) = \min\Big\{\#\chi(D) \;\Big|\; D \text{ is an } [n,r;q] \text{ subcode of } C\Big\}.$$

In particular, the minimum distance of C is d_1. The *weight hierarchy* of C is the set $\{d_1, d_2, \cdots, d_k\}$. The weight hierarchy satisfies the following inequality:

$$d_r \ge d_{r-1}\left(1 + \frac{q-1}{q^r - q}\right). \tag{1.17}$$

In particular, we have

$$d_r \ge d_{r-1} + 1. \tag{1.18}$$

An upper bound that follows from (1.18) is the *generalized Singleton bound*

$$d_r \le n - k + r. \tag{1.19}$$

1.6 Principles of error detection

1.6.1 *Pure detection*

Consider what happens when a code word $\mathbf{x}$ from an (n,M) code C is transmitted over a channel K and errors occur during transmission. If the received vector $\mathbf{y}$ is not a code word we immediately realize that something has gone wrong during transmission, we *detect* that errors have occurred. However, it may happen that the combination of errors is such that the received vector $\mathbf{y}$ is also a code word. In this case we have no way to tell that the received code word is not the sent code word. Therefore, we have an *undetectable error.* We let $P_{\text{ue}} = P_{\text{ue}}(C,K)$ denote the probability that this happens. It is called the *probability of undetected error.* If $P(\mathbf{x})$ is the probability that $\mathbf{x}$ was sent and $P(\mathbf{y}|\mathbf{x})$ is the probability that $\mathbf{y}$ is received, given that $\mathbf{x}$ was sent, then

$$P_{\text{ue}}(C,K) = \sum_{\mathbf{x}\in C} P(\mathbf{x}) \sum_{\mathbf{y}\in C\setminus\{\mathbf{x}\}} P(\mathbf{y}|\mathbf{x}).$$

In most cases we will assume that each code word is equally likely to be sent, that is, $P(\mathbf{x}) = \frac{1}{M}$. Under this assumption we get

$$P_{\mathrm{ue}}(C,K) = \frac{1}{M} \sum_{\mathbf{x}\in C} \sum_{\mathbf{y}\in C\setminus\{\mathbf{x}\}} P(\mathbf{y}|\mathbf{x}).$$

The quantity $P_{\mathrm{ue}}(C,K)$ is a main parameter for describing how well C performs on the channel K, and it is the main subject of study in this book. In Chapter 2, we study $P_{\mathrm{ue}}(C,K)$ for the q-ary symmetric channel, in Chapter 3 we describe results that are particular for the binary symmetric channel, in Chapter 4 we study other channels.

Remark 1.1. It is easy to show that for any channel K with additive noise and any coset S of a linear code C we have $P_{\mathrm{ue}}(C,K) = P_{\mathrm{ue}}(S,K)$.

1.6.2 *Combined correction and detection*

In some applications we prefer to use some of the power of a code to correct errors and the remaining power to detect errors. Suppose that C is an $(n,M;q)$ code capable of correcting all error patterns with t_0 or less errors that can occur on the channel and suppose that we use the code to correct all error patterns with t errors or less, where $t \le t_0$. Let $M_t(\mathbf{x})$ be the set of all vectors $\mathbf{y}$ such that $d_{\mathrm{H}}(\mathbf{x},\mathbf{y}) \le t$ and such that $\mathbf{y}$ can be received when $\mathbf{x}$ is sent over the channel. For two distinct $\mathbf{x}_1, \mathbf{x}_2 \in C$, the sets $M_t(\mathbf{x}_1)$, $M_t(\mathbf{x}_2)$ are disjoint. If $\mathbf{y} \in M_t(\mathbf{x})$ is received, we decode into $\mathbf{x}$. If $\mathbf{y} \notin M_t(\mathbf{x})$ for all $\mathbf{x} \in C$, then we detect an error.

Suppose that $\mathbf{x}$ is sent and $\mathbf{y}$ is received. There are then three possibilities:

(1) $\mathbf{y} \in M_t(\mathbf{x})$. We then decode, correctly, into $\mathbf{x}$.
(2) $\mathbf{y} \notin M_t(\mathbf{x}')$ for all $\mathbf{x}' \in C$. We then detect an error.
(3) $\mathbf{y} \in M_t(\mathbf{x}')$ for some $\mathbf{x}' \in C \setminus \{\mathbf{x}\}$. We then decode erroneously into $\mathbf{x}'$, and we have an undetectable error.

Let $P_{\mathrm{ue}}^{(t)} = P_{\mathrm{ue}}^{(t)}(C,K)$ denote the probability that we have an undetectable error. As above we get

$$P_{\mathrm{ue}}^{(t)}(C,K) = \sum_{\mathbf{x}\in C} P(\mathbf{x}) \sum_{\mathbf{x}'\in C\setminus\{\mathbf{x}\}} \sum_{\mathbf{y}\in M_t(\mathbf{x}')} P(\mathbf{y}|\mathbf{x}).$$

Assuming that $P(\mathbf{x}) = \frac{1}{M}$ for all $\mathbf{x}' \in C$, we get

$$P_{\mathrm{ue}}^{(t)}(C,K) = \frac{1}{M} \sum_{\mathbf{x}\in C} \sum_{\mathbf{x}'\in C\setminus\{\mathbf{x}\}} \sum_{\mathbf{y}\in M_t(\mathbf{x}')} P(\mathbf{y}|\mathbf{x}).$$

1.7 Comments and references

1.1 Most of this material can be found in most text books on error-correcting codes, see the general bibliography. However, many of the books restrict themselves to binary codes.

1.2 Again, this is mainly standard material.

1.3 Some of this material is standard. Most textbooks restrict their presentation to linear codes and, therefore, to the weight distribution.

Theorem 1.3 is due to Pless (1963).

Theorems 1.5 and 1.6 are due to Delsarte (1972).

Binomial moments seems to have been used for the first time by MacWilliams (1963). Possibly the first application to error detection is by Kløve (1984d). A survey on binomial moments was given by Dodunekova (2003b).

Theorem 1.9 and Corollary 1.2 were given in Kløve and Korzhik (1995, pp. 51–52) in the binary case. For general q, they were given by Dodunekova (2003b).

Theorems 1.10 and 1.11 is due to AbdelGhaffar (1997).

Theorem 1.12 is essentially due to AbdelGhaffar (2004). Corollary 1.3 (for $q = 2$) was first given by Fu, Kløve, and Wei (2003), with a different proof.

1.4 Theorem 1.14 is due to MacWilliams (1963). Theorem 1.15 (for $q = 2$) was given by Kasami, Fujiwara, and Lin (1986).

Theorem 1.16 is from Kløve (1992).

Theorem 1.17 and Corollary 1.8 are due to Assmus and Mattson (1978).

Theorem 1.18 is essentially due to Ancheta (1981).

Theorem 1.19 with $q = 2$ is due to Sullivan (1967). An alternative proof and generalization to general q was given by Redinbo (1973). Further results are given in Kløve (1993), Kløve (1994b), Kløve (1996c).

We remark that the weight distribution of cosets can be useful in the wire-tap channel area, see Wyner (1975) and Korzhik and Yakovlev (1992).

Theorem 1.21 is essentially due to MacWilliams (1963). In the present form it was given in Kløve (1984a).

Theorem 1.22 was given in Kløve and Korzhik (1995, Section 2.2). Special cases were given implicitly in Korzhik and Fink (1975) and Kasami, Kløve, and Lin (1983).

Theorem 1.23 is due to Kløve (1996a).

The weight hierarchy (under a different name) was first studied by

Helleseth, Kløve, and Mykkeltveit (1977). The r-th minimum support weight is also known as *r-th generalized Hamming weight*, see Wei (1991). The inequality (1.17) was shown by Helleseth, Kløve, and Ytrehus (1992) (for $q = 2$) and Helleseth, Kløve, and Ytrehus (1993) (for general q).

1.6. A more detailed discussion of combined error detection and correction is found for example in Kløve (1984a).

Chapter 2

Error detecting codes for the q-ary symmetric channel

The q-ary symmetric channel (qSC) is central in many applications and we will therefore give a fairly complete account of the known results. Results that are valid for all q are given in this chapter. A special, important case is the binary case ($q = 2$). Results that are particular to the binary case will be given in the next chapter.

2.1 Basic formulas and bounds

2.1.1 *The q-ary symmetric channel*

The q-ary symmetric channel (qSC) with error probability parameter p is defined by the transition probabilities

$$P(b|a) = \begin{cases} 1 - p \text{ if } b = a, \\ \frac{p}{q-1} \text{ if } b \neq a. \end{cases}$$

The parameter p is known as the *symbol error probability.*

2.1.2 *Probability of undetected error*

Suppose $\mathbf{x} \in F_q^n$ is sent over the q-ary symmetric channel with symbol error probability p, that errors are independent, and that $\mathbf{y}$ received. Since exactly $d_{\mathrm{H}}(\mathbf{x}, \mathbf{y})$ symbols have been changed during transmission, the remaining $n - d_{\mathrm{H}}(\mathbf{x}, \mathbf{y})$ symbols are unchanged, and we get

$$P(\mathbf{y}|\mathbf{x}) = \left(\frac{p}{q-1}\right)^{d_{\mathrm{H}}(\mathbf{x},\mathbf{y})} (1-p)^{n-d_{\mathrm{H}}(\mathbf{x},\mathbf{y})}.$$

Assume that C is a code over F_q of length n and that the code words are equally likely to be chosen for transmission over qSC. For this situation, we will use the notation $P_{\text{ue}}(C,p) = P_{\text{ue}}(C,qSC)$ for the probability of undetected error. It is the main subject of study in this chapter.

If $A_C(z)$ denotes the distance distribution function of C, then

$$\begin{aligned}P_{\text{ue}}(C,p) &= \frac{1}{M}\sum_{\mathbf{x}\in C}\sum_{\mathbf{y}\in C\setminus\{\mathbf{x}\}}\Big(\frac{p}{q-1}\Big)^{d_{\text{H}}(\mathbf{x},\mathbf{y})}(1-p)^{n-d_{\text{H}}(\mathbf{x},\mathbf{y})}\\ &= \sum_{i=1}^{n} A_i(C)\Big(\frac{p}{q-1}\Big)^i(1-p)^{n-i}\\ &= (1-p)^n\sum_{i=1}^{n} A_i(C)\Big(\frac{p}{(q-1)(1-p)}\Big)^i\\ &= (1-p)^n\Big\{A_C\Big(\frac{p}{(q-1)(1-p)}\Big)-1\Big\}.\end{aligned}$$

We state this basic result as a theorem.

Theorem 2.1. *Let C be an $(n,M;q)$ code. Then*

$$\begin{aligned}P_{\text{ue}}(C,p) &= \frac{1}{M}\sum_{\mathbf{x}\in C}\sum_{\mathbf{y}\in C\setminus\{\mathbf{x}\}}\Big(\frac{p}{q-1}\Big)^{d_{\text{H}}(\mathbf{x},\mathbf{y})}(1-p)^{n-d_{\text{H}}(\mathbf{x},\mathbf{y})}\\ &= \sum_{i=1}^{n} A_i(C)\Big(\frac{p}{q-1}\Big)^i(1-p)^{n-i}\\ &= (1-p)^n\Big\{A_C\Big(\frac{p}{(q-1)(1-p)}\Big)-1\Big\}.\end{aligned}$$

An $(n,M;q)$ code C is called *optimal* (error detecting) for p if $P_{\text{ue}}(C,p) \le P_{\text{ue}}(C',p)$ for all $(n,M;q)$ codes C'. Similarly, an $[n,k;q]$ code is called an *optimal linear code* for p if $P_{\text{ue}}(C,p) \le P_{\text{ue}}(C',p)$ for all $[n,k;q]$ codes C'. Note that a linear code may be an optimal linear without being optimal over all codes. However, to simplify the language, we talk about *optimal codes*, meaning optimal in the general sense if the code is non-linear and optimal among linear codes if the code is linear.

When we want to find an $(n,M;q)$ or $[n,k;q]$ code for error detection in some application, the best choice is an optimal code for p. There are two problems. First, we may not know p, and a code optimal for $p' \ne p$ may not be optimal for p. Moreover, even if we know p, there is in general no method to find an optimal code, except exhaustive search, and this is in most cases not feasible. Therefore, it is useful to have some criterion by which we can judge the usefulness of a given code for error detection.

We note that $P_{\text{ue}}\Big(C,\frac{q-1}{q}\Big) = \frac{M-1}{q^n}$. It used to be believed that since $p = \frac{q-1}{q}$ is the "worst case", it would be true that $P_{\text{ue}}(C,p) \le \frac{M-1}{q^n}$ for all

$p \in [0, \frac{q-1}{q}]$. However, this is not the case as shown by the following simple example.

Example 2.1. Let $C = \{(a, b, 0) \mid a, b \in F_q\}$. It is easy to see that for each code word $\mathbf{c} \in C$ there are $2(q-1)$ code words in C at distance one and $(q-1)^2$ code words at distance 2. Hence

$$A_0 = 1,\ A_1 = 2(q-1),\ A_2 = (q-1)^2,$$

and

$$P_{\text{ue}}(C,p) = 2(q-1)\frac{p}{q-1}(1-p)^2 + (q-1)^2\Big(\frac{p}{q-1}\Big)^2(1-p) = 2p(1-p)^2 + p^2(1-p).$$

This function takes it maximum in $[0, \frac{q-1}{q}]$ for $p = 1 - \frac{1}{\sqrt{3}}$. In particular,

$$P_{\text{ue}}\Big(C, 1 - \frac{1}{\sqrt{3}}\Big) = \frac{2}{3\sqrt{3}} \approx 0.3849 > \frac{q^2-1}{q^3} = P_{\text{ue}}\Big(C, \frac{q-1}{q}\Big)$$

for all $q \geq 2$.

In fact, $P_{\text{ue}}(C,p)$ may have more than one local maximum in the interval $[0, (q-1)/q]$.

Example 2.2.
Let C be the $(13, 21; 2)$ code given in Table 2.1.

Table 2.1 Code in Example 2.2.

(1111111111110)	(1111000000000)	(1100110000000)	(1100001100000)
(1100000011000)	(1100000000110)	(0011110000000)	(0011001100000)
(0011000011000)	(0011000000110)	(0000111100000)	(0000110011000)
(0000110000110)	(0000001111000)	(0000001100110)	
(1010101011101)	(0101011001000)	(1010101010101)	
(1010010110011)	(1001100101011)	(0101100110101)	

The distance distribution of C is given in Table 2.2.

Table 2.2 Distance distribution for the code in Example 2.2.

i	1	2	3	4	5	6	7	8	9	10	11	12	13
$210A_i$	1	0	0	52	10	9	68	67	1	2	0	0	0

The probability of undetected error for this code has *three* local maxima in the interval $[0, 1/2]$, namely for $p = 0.0872$, $p = 0.383$, and $p = 0.407$.

An $(n, M; q)$ code C is called *good* (for error detection) if

$$P_{\text{ue}}(C,p) \le P_{\text{ue}}\Big(C, \frac{q-1}{q}\Big) = \frac{M-1}{q^n} \tag{2.1}$$

for all $p \in [0, \frac{q-1}{q}]$. Note that "good" is a technical and relative term. An extreme case is the code F_q^n which cannot detect any errors. Since $P_{\text{ue}}(F_q^n, p) = 0$ for all p, the code is "good" in the sense defined above even if it cannot detect any errors!

An engineering rule of thumb is that if a code, *with acceptable parameters* (length and size), is good in the sense just defined, then it is good enough for most practical applications. It has not been proved that there exist good $(n, M; q)$ codes for all n and $M \le q^n$, but numerical evidence indicates that this may be the case.

We shall later show that a number of well known classes of codes are good. On the other hand, many codes are not good. Therefore, it is important to have methods to decide if a code is good or not.

A code which is not good is called *bad*, that is, a code C is bad if $P_{\text{ue}}(C,p) > \frac{M-1}{q^n}$ for some $p \in [0, \frac{q-1}{q}]$. If C satisfy the condition $P_{\text{ue}}(C,p) \le \frac{M}{q^n}$ for all $p \in [0, \frac{q-1}{q}]$, we call it *satisfactory*. Clearly, "satisfactory" is a weaker condition than "good". A code that is not satisfactory is called *ugly*, that is, a code C is ugly if $P_{\text{ue}}(C,p) > \frac{M}{q^n}$ for some $p \in [0, \frac{q-1}{q}]$. Some authors use the term *good* for codes which are called *satisfactory* here.

The bound $\frac{M}{q^n}$ in the definition of a satisfactory code is to some extent arbitrary. For most practical applications, any bound of the same order of magnitude would do. Let $\mathcal{C}$ be an infinite class of codes. We say that $\mathcal{C}$ is *asymptotically good* if there exists a constant c such that

$$P_{\text{ue}}(C,p) \le c P_{\text{ue}}\Big(C, \frac{q-1}{q}\Big)$$

for all $C \in \mathcal{C}$ and all $p \in \left[0, \frac{q-1}{q}\right]$. Otherwise we say that $\mathcal{C}$ is *asymptotically bad.*

A code C is called *proper* if $P_{\text{ue}}(C,p)$ is monotonously increasing on $[0, \frac{q-1}{q}]$. A proper code is clearly good, but a code may be good without being proper.

A simple, but useful observation is the following lemma.

Lemma 2.1. *For $i \le j$ and $p \in \left[0, \frac{q-1}{q}\right]$, we have*

$$\Big(\frac{p}{q-1}\Big)^i (1-p)^{n-i} \ge \Big(\frac{p}{q-1}\Big)^j (1-p)^{n-j}. \tag{2.2}$$

Proof. We note that (2.2) is equivalent to

$$\Big(\frac{p}{(q-1)(1-p)}\Big)^i \geq \Big(\frac{p}{(q-1)(1-p)}\Big)^j,$$

and this is satisfied since

$$\frac{p}{(q-1)(1-p)} \leq 1.$$

□

When we want to compare the probability of undetected error for two codes, the following lemma is sometimes useful.

Lemma 2.2. *Let $x_1, x_2, \ldots x_n$ and $\gamma_1, \gamma_2, \ldots, \gamma_n$ be real numbers such that*

$$x_1 \geq x_2 \geq \cdots \geq x_n \geq 0$$

and

$$\sum_{i=1}^{j} \gamma_i \geq 0 \text{ for } j = 1, 2, \ldots, n.$$

Then

$$\sum_{i=1}^{n} \gamma_i x_i \geq 0.$$

Proof. Let $\sigma_j = \gamma_1 + \gamma_2 + \cdots + \gamma_j$. In particular, $\sigma_0 = 0$ and by assumption, $\sigma_j \geq 0$ for all j. Then

$$\begin{aligned}\sum_{i=1}^{n} \gamma_i x_i &= \sum_{i=1}^{n} (\sigma_i - \sigma_{i-1}) x_i = \sum_{i=1}^{n} \sigma_i x_i - \sum_{i=0}^{n-1} \sigma_i x_{i+1} \\ &= \sigma_n x_n + \sum_{i=1}^{n-1} \sigma_i (x_i - x_{i+1}) \geq 0.\end{aligned}$$

□

Corollary 2.1. *If C and C' are $(n, M; q)$ codes such that*

$$\sum_{i=1}^{j} A_i(C) \leq \sum_{i=1}^{j} A_i(C')$$

for all $j = 1, 2, \ldots, n$, then

$$P_{\text{ue}}(C, p) \leq P_{\text{ue}}(C', p)$$

for all $p \in [0, (q-1)/q]$.

Proof. The results follows from Lemma 2.2 choosing $\gamma_i = A_i(C') - A_i(C)$ and $x_i = \left(\frac{p}{q-1}\right)^i (1-p)^{n-i}$. Lemma 2.1 shows that the first condition in Lemma 2.2 is satisfied; the second condition is satisfied by assumption. □

Example 2.3. We consider the possible distance distributions of $(5,4;2)$ codes. There are 38 different distance distributions of $(5,4;2)$ codes; of these 10 occur for linear codes. It turns out that $2A_i$ is always an integer. Therefore, we list those values in two tables, Table 2.3 for weight distributions of linear codes (in some cases there exist non-linear codes also with these distance distributions) and Tables 2.4 for distance distributions which occur only for non-linear codes.

Table 2.3 Possible weight distributions for linear $[5,2;2]$ codes.

$2A_1$	$2A_2$	$2A_3$	$2A_4$	$2A_5$	type[a]	no. of nonlinear	no. of linear
0	0	4	2	0	P	0	3
0	2	0	4	0	P	0	2
0	2	2	0	2	P	0	2
0	2	4	0	0	P	12	6
0	4	0	2	0	P	24	3
0	6	0	0	0	S	4	2
2	0	0	2	2	G	0	1
2	0	2	2	0	G	0	4
2	2	2	0	0	S	24	6
4	2	0	0	0	U	0	2

[a]P: proper, G: good, but not proper, S: satisfactory, but bad, U: ugly.

We note that if C is a $(5,4;2)$ code, and we define C' by taking the cyclic shift of each code word, that is

$$C' = \{(c_5, c_1, c_2, c_3, c_4) \mid (c_1, c_2, c_3, c_4, c_5) \in C\},$$

then C' and C have the same distance distribution. Moreover, the five codes obtained by repeating this cycling process are all distinct. Hence, the codes appear in groups of five equivalent code. In the table, we have listed the number of such groups of codes with a given weight distribution (under the headings "no. of nonlinear" and "no. of linear").

Using Corollary 2.1, it is easy to see that $P_{\text{ue}}(C,p) \leq P_{\text{ue}}(C',p)$ for all $(5,4;2)$ codes C and all $p \in [0,1/2]$, where C' is the linear $[5,2;2]$ code with weight distribution $(1,2,1,0,0,0)$. A slightly more complicated argument shows that $P_{\text{ue}}(C,p) \geq P_{\text{ue}}(C'',p)$ for all $(5,4;2)$ codes C and all

Table 2.4 The other distance distributions for $(5, 4; 2)$ codes.

$2A_1$	$2A_2$	$2A_3$	$2A_4$	$2A_5$	type[a]	no. of nonlinear
0	1	3	2	0	P	24
0	1	4	1	0	P	24
0	2	2	1	1	P	12
0	2	3	0	1	P	24
0	2	3	1	0	P	48
0	3	0	3	0	P	32
0	3	2	0	1	P	24
0	3	3	0	0	P	24
0	5	0	1	0	S	24
1	0	3	2	0	S	24
1	1	1	2	1	P	16
1	1	2	1	1	P	24
1	1	2	2	0	P	72
1	1	3	1	0	P	48
1	2	1	1	1	P	24
1	2	2	0	1	P	12
1	2	2	1	0	P	60
1	2	3	0	0	S	48
1	3	2	0	0	S	48
2	0	1	2	1	G	8
2	1	0	2	1	S	8
2	1	1	1	1	S	16
2	1	1	2	0	S	8
2	1	2	1	0	S	48
2	2	1	1	0	S	48
2	3	1	0	0	S	24
3	2	1	0	0	U	24
3	3	0	0	0	U	8

[a]P: proper, G: good, but not proper,
S: satisfactory, but bad, U: ugly.

$p \in [0, 1/2]$, where C'' is the linear $[5, 2; 2]$ code with weight distribution $(1, 0, 0, 2, 1, 0)$.

For a practical application we may know that $p \leq p_0$ for some fixed p_0. If we use an $(n, M, d; q)$ code with $d \geq p_0 n$, then the next theorem shows that $P_{ue}(p) \leq P_{ue}(p_0)$ for all $p \leq p_0$.

Theorem 2.2. *Let C be an $(n, M, d; q)$ code. Then $P_{ue}(C, p)$ is monotonously increasing on $\left[0, \frac{d}{n}\right]$.*

Proof. Since $p^i(1-p)^{n-i}$ is monotonously increasing on $\left[0, \frac{i}{n}\right]$, and in particular on $\left[0, \frac{d}{n}\right]$ for all $i \geq d$, the theorem follows. □

2.1.3 *The threshold*

Many codes are not good for error detection (in the technical sense). On the other hand, p is usually small in most practical applications and (2.1) may well be satisfied for the actual values of p. Therefore, we consider the *threshold* of C, which is defined by

$$\theta(C) = \max\left\{p' \in \left[0, \frac{q-1}{q}\right] \;\middle|\; P_{ud}(C,p) \leq P_{ud}\Big(C, \frac{q-1}{q}\Big) \text{ for all } p \in [0,p']\right\}. \tag{2.3}$$

For $p \leq \theta(C)$ the bound (2.1) is still valid. In particular, C is good for error detection if and only if $\theta(C) = (q-1)/q$. Note that $\theta(C)$ is a root of the equation $P_{ud}(C,p) = P_{ud}(C,(q-1)/q)$, and it is the smallest root in the interval $(0,(q-1)/q]$, except in the rare cases when $P_{ud}(C,p)$ happens to have a local maximum for this smallest root. To determine the threshold exactly is difficult in most cases and therefore it is useful to have estimates.

Theorem 2.3. *Let $\psi(\delta;q)$ be the least positive root of the equation*

$$\Big(\frac{\psi}{q-1}\Big)^{\delta}(1-\psi)^{1-\delta} = \frac{1}{q}.$$

If C is an $(n,M,d;q)$ code, then $\theta(C) > \psi(d/n;q)$.

Proof. For $p \leq \psi = \psi(d/n;q)$ we have, by Lemma 2.1,

$$\begin{aligned}
P_{ud}(C,p) &= \sum_{i=d}^{n} A_i\Big(\frac{p}{q-1}\Big)^{i}(1-p)^{n-i} \\
&\leq \Big(\frac{p}{q-1}\Big)^{d}(1-p)^{n-d}\sum_{i=d}^{n} A_i \\
&= \bigg(\Big(\frac{p}{q-1}\Big)^{\delta}(1-p)^{1-\delta}\bigg)^{n}(M-1) \\
&\leq \bigg(\Big(\frac{\psi}{q-1}\Big)^{\delta}(1-\psi)^{1-\delta}\bigg)^{n}(M-1) \\
&= \frac{1}{q^n}(M-1) = P_{ud}\Big(C, \frac{q-1}{q}\Big).
\end{aligned}$$

Hence $\theta(C) > \psi$. □

Example 2.4. For $m \geq 1$, let C_m be the binary code generated by the matrix

$$\begin{bmatrix} \overbrace{1\ldots1}^{m} & \overbrace{0\ldots0}^{m} & \overbrace{0\ldots0}^{m} \\ 0\ldots0 & 1\ldots1 & 0\ldots0 \\ 0\ldots0 & 0\ldots0 & 1\ldots1 \end{bmatrix}.$$

Clearly, C_m is a $[3m, 3, m; 2]$ code and

$$P_{\text{ue}}(C_m, p) = 3p^m(1-p)^{2m} + 3p^{2m}(1-p)^m + p^{3m}.$$

For $m \leq 3$, C_m is proper. The code C_4 is good, but not proper. For the codes C_m, $d/n = 1/3$. For $m \geq 5$, we have

$$\frac{P_{\text{ue}}(C_m, 1/3)}{P_{\text{ue}}(C_m, 1/2)} \geq \frac{3(4/27)^m}{7/8^m} = \frac{3}{7}\Big(\frac{32}{27}\Big)^m > 1,$$

and the code C_m is bad.

We have

$$\psi\Big(\frac{1}{3};2\Big)\Big(1-\psi\Big(\frac{1}{3};2\Big)\Big)^2 = \frac{1}{2}$$

and so $\psi(1/3;2) \approx 0.190983$. Hence $\theta(C_m) > 0.190983$. On the other hand, if σ_m is the least positive root of $3\sigma^m(1-\sigma)^{2m} = 7 \cdot 2^{-3m}$, that is

$$\sigma(1-\sigma)^2 = \Big(\frac{3}{7}\Big)^{1/m}\frac{1}{2},$$

then

$$P_{\text{ue}}(C, \sigma) = 3\sigma^m(1-\sigma)^{2m} + 3\sigma^{2m}(1-\sigma)^m + \sigma^{3m} > 7 \cdot 2^{-3m} = P_{\text{ue}}\Big(C, \frac{1}{2}\Big)$$

and so $\theta(C_m) < \sigma_m$. Since $(3/7)^{1/m} \to 1$ when $m \to \infty$, we see that $\sigma_m \to \psi(1/3, 2)$. In Table 2.5 we give the numerical values in some cases. The values illustrate that σ_m is a good approximation to $\theta(C_m)$.

Table 2.5 Selected values for Example 2.4.

m	$\theta(C_m)$	σ_m
5	0.3092	0.3253
6	0.27011	0.27055
10	0.22869306	0.22869335
30	0.201829421660768430283	0.201829421660768430299

2.1.4 *Alternative expressions for the probability of undetected error*

There are several alternative expressions for $P_{ue}(C,p)$. Using Lemma 1.1, we get

$$A_C\Big(\frac{p}{(q-1)(1-p)}\Big) = \frac{M}{q^n}(1-p)^{-n}A_C^{\perp}\Big(1-\frac{qp}{q-1}\Big)$$

and so Theorem 2.1 implies the following result.

Theorem 2.4. *Let C be an $(n,M;q)$ code. Then*

$$P_{ue}(C,p) = \frac{M}{q^n}A_C^{\perp}\Big(1-\frac{qp}{q-1}\Big) - (1-p)^n.$$

In particular, if C is a linear $[n,k;q]$ code, then

$$\begin{aligned}P_{ue}(C,p) &= q^{k-n}A_{C^{\perp}}\Big(1-\frac{qp}{q-1}\Big) - (1-p)^n\\ &= q^{k-n}\sum_{i=0}^{n} A_i(C^{\perp})\Big(1-\frac{qp}{q-1}\Big)^i - (1-p)^n.\end{aligned}$$

Example 2.5. As an illustration, we show that Hamming codes are proper. The $\Big[n = \frac{q^m-1}{q-1}, k = \frac{q^m-1}{q-1} - m; q\Big]$ Hamming code C is the dual of the $\Big[\frac{q^m-1}{q-1}, m; q\Big]$ Simplex code S_m. All the non-zero code words in S_m have weight $q^{m-1} = (n(q-1)+1)/q$ and $q^m - 1 = n(q-1)$. Since $q^{k-n} = q^{-m} = 1/(n(q-1)+1)$, we get

$$P_{ue}(C,p) = \frac{1}{n(q-1)+1}\Big(1 + n(q-1)\Big(1-\frac{q}{q-1}p\Big)^{\frac{n(q-1)+1}{q}}\Big) - (1-p)^n.$$

Hence for $p \in \Big(0, \frac{q-1}{q}\Big]$ we get

$$\begin{aligned}\frac{d}{dp}P_{ue}(C,p) &= -n\Big(1-\frac{q}{q-1}p\Big)^{\frac{n(q-1)+1}{q}-1} + n(1-p)^{n-1}\\ &= n\Big\{\Big((1-p)^{q/(q-1)}\Big)^{\frac{(n-1)(q-1)}{q}} - \Big(1-\frac{q}{q-1}p\Big)^{\frac{(n-1)(q-1)}{q}}\Big\} > 0\end{aligned}$$

since

$$(1-p)^{q/(q-1)} > \Big(1-\frac{q}{q-1}p\Big).$$

Therefore, C is proper.

We go on to give one more useful expression for $P_{\rm ue}(C,p)$.

Theorem 2.5. *Let C be an $(n, M; q)$ code. Then*

$$P_{\rm ue}(C,p) = \sum_{i=1}^{n} A_i^{\diamond}(C)\Big(\frac{p}{q-1}\Big)^i\Big(1-\frac{q}{q-1}p\Big)^{n-i}.$$

Proof. Combining Theorems 1.7 and 2.1, we get

$$\begin{aligned} P_{\rm ue}(C,p) &= (1-p)^n\Big\{A_C\Big(\frac{p}{(q-1)(1-p)}\Big)-1\Big\} \\ &= (1-p)^n\sum_{i=1}^{n} A_i^{\diamond}(C)\Big(\frac{p}{(q-1)(1-p)}\Big)^i\Big(1-\frac{p}{(q-1)(1-p)}\Big)^{n-i} \\ &= \sum_{i=1}^{n} A_i^{\diamond}(C)\Big(\frac{p}{q-1}\Big)^i\Big(1-\frac{q}{q-1}p\Big)^{n-i}. \end{aligned}$$

□

2.1.5 *Relations to coset weight distributions*

We here also mention another result that has some applications. The result follows directly from Theorem 1.19.

Theorem 2.6. *Let C be an $[n, k; q]$ code and S a proper coset of C. Let $\mathbf{0}$ be sent over qSC, and let $\mathbf{y}$ be the received vector. Then*

$$\frac{Pr(\mathbf{y}\in S)}{Pr(\mathbf{y}\in C)} = \frac{(1-p)^n A_S^w\Big(\frac{p}{(q-1)(1-p)}\Big)}{(1-p)^n A_C\Big(\frac{p}{(q-1)(1-p)}\Big)} \le \frac{1-\Big(1-\frac{qp}{q-1}\Big)^{k+1}}{1+(q-1)\Big(\frac{qp}{q-1}\Big)^{k+1}}.$$

2.2 $P_{\rm ue}$ for a code and its MacWilliams transform

Let C be an $(n, M; q)$ code. Define $P_{\rm ue}^{\perp}(C,p)$ by

$$P_{\rm ue}^{\perp}(C,p) = \sum_{i=1}^{n} A_i^{\perp}(C)\Big(\frac{p}{q-1}\Big)^i(1-p)^{n-i}.$$

If C is linear, then Theorem 1.14 implies that $P_{\rm ue}^{\perp}(C,p) = P_{\rm ue}(C^{\perp},p)$. Similarly to Theorem 2.4 we get

$$P_{\rm ue}^{\perp}(C,p) = \frac{1}{M}A_C\Big(1-\frac{qp}{q-1}\Big)-(1-p)^n. \tag{2.4}$$

Theorem 2.7. *Let C be an $(n, M; q)$ code. Then*

$$P_{\rm ue}(C,p) = M(1-p)^n P_{\rm ue}^{\perp}\Big(C,\frac{q-1-qp}{q-qp}\Big)+\frac{M}{q^n}-(1-p)^n$$

and

$$P_{\rm ue}^{\perp}(C,p) = \frac{q^n}{M}(1-p)^n P_{\rm ue}\Big(C, \frac{q-1-qp}{q-qp}\Big) + \frac{1}{M} - (1-p)^n.$$

Proof. From (2.4) we get

$$\begin{aligned} P_{\rm ue}(C,p) &= (1-p)^n\Big\{A_C\Big(\frac{p}{(q-1)(1-p)}\Big) - 1\Big\} \\ &= (1-p)^n\Big\{A_C\Big(1 - \frac{q}{q-1}\cdot\frac{q-1-qp}{q-qp}\Big) - 1\Big\} \\ &= (1-p)^n\Big\{MP_{\rm ue}^{\perp}\Big(C, \frac{q-1-qp}{q-qp}\Big) + M\Big(1 - \frac{q-1-qp}{q-qp}\Big)^n - 1\Big\} \\ &= M(1-p)^n P_{\rm ue}^{\perp}\Big(C, \frac{q-1-qp}{q-qp}\Big) + \frac{M}{q^n} - (1-p)^n. \end{aligned}$$

The proof of the other relation is similar. □

The dual of a proper linear code may not be proper, or even good, as shown by the next example.

Example 2.6. Consider the code C_5 defined in Example 2.4. It was shown in that example that C_5 is bad. On the other hand,

$$P_{\rm ue}(C_5^{\perp}, p) = \frac{1}{8}\Big\{1 + 3(1-2p)^5 + 3(1-2p)^{10} + (1-2p)^{15}\Big\} - (1-p)^{15},$$

and this is increasing on $[0, 1/2]$. Hence $C_5^{\perp}$ is proper, but the dual is bad.

There are a couple of similar conditions, however, such that if $P_{\rm ue}(C,p)$ satisfies the condition, then so does $P_{\rm ue}^{\perp}(C,p)$.

Theorem 2.8. *Let C be an $(n, M; q)$ code.*

(1) $P_{\rm ue}(C,p) \le Mq^{-n}$ *for all* $p \in \left[0, \frac{q-1}{q}\right]$,
if and only if $P_{\rm ue}^{\perp}(C,p) \le 1/M$ *for all* $p \in \left[0, \frac{q-1}{q}\right]$.

(2) $P_{\rm ue}(C,p) \le Mq^{-n}\Big\{1 - (1-p)^k\Big\}$ *for all* $p \in \left[0, \frac{q-1}{q}\right]$,
if and only if $P_{\rm ue}^{\perp}(C,p) \le 1/M\Big\{1 - (1-p)^{n-k}\Big\}$ *for all* $p \in \left[0, \frac{q-1}{q}\right]$.

(3) $P_{\rm ue}(C,p) \le \frac{M-1}{q^n-1}\Big\{1 - (1-p)^n\Big\}$ *for all* $p \in \left[0, \frac{q-1}{q}\right]$,
if and only if $P_{\rm ue}^{\perp}(C,p) \le \frac{q^n/M-1}{q^n-1}\Big\{1 - (1-p)^n\Big\}$ *for all* $p \in \left[0, \frac{q-1}{q}\right]$.

Proof. Assume that $P_{\rm ue}(C,p) \le Mq^{-n}$ for all $p \in \left[0, \frac{q-1}{q}\right]$. Using Theorem 2.7 we see that if $P_{\rm ue}(C,p) \le Mq^{-n}$, then

$$\begin{aligned} P_{\rm ue}^{\perp}(C^{\perp},p) &= \frac{q^n}{M}(1-p)^n P_{\rm ue}\Big(C, \frac{q-1-qp}{q-qp}\Big) + \frac{1}{M} - (1-p)^n \\ &\le \frac{q^n}{M}(1-p)^n \frac{M}{q^n} + \frac{1}{M} - (1-p)^n = \frac{1}{M}. \end{aligned}$$

The proof of the other relations are similar. □

Note that for a linear code C, the relation (1) is equivalent to the statement that C is satisfactory if and only if $C^{\perp}$ is satisfactory.

2.3 Conditions for a code to be satisfactory, good, or proper

2.3.1 *How to determine if a polynomial has a zero*

In a number of cases we want to know if a polynomial has a zero in a given interval. For example, if $\frac{d}{dp} P_{\rm ue}(C,p) > 0$ for all $p \in \left(0, \frac{q-1}{q}\right)$, then C is proper. If $P_{\rm ue}\left(C, \frac{q-1}{q}\right) - P_{\rm ue}(C,p) \ge 0$ then C is good (by definition). If $A_{C_1}(z) - A_{C_2}(z) > 0$ for $z \in (0,1)$, then $P_{\rm ue}(C_1,p) > P_{\rm ue}(C_2,p)$ for all $p \in \left(0, \frac{q-1}{q}\right)$, and so C_2 is better for error detection than C_1. The simplest way is to use some mathematical software to draw the graph and inspect it and/or calculate the roots. However, this may sometimes be misleading. Therefore, we give a short description of the systematic method of Sturm sequences.

Let $f(z)$ be a polynomial and $[a,b]$ an interval.

Define the sequence $f_0(z), f_1(z), \cdots, f_m(z)$ by the Euclidean algorithm:

$f_0(z) = f(z)$,
$f_1(z) = f'(z)$,
$f_i(z) \equiv -f_{i-2}(z) \pmod{f_{i-1}(z)}$, $\deg(f_i(z)) < \deg(f_{i-1}(z))$ for $2 \le i \le m$,
$f_m(z)$ divides $f_{m-1}(z)$.

Define $g_0(z), g_1(z), \cdots, g_m(z)$ by

$$g_i(z) = \frac{f_i(z)}{f_m(z)}.$$

If $g_0(a) = 0$ we divide all $g_i(z)$ with $(z-a)$; and if $g_0(b) = 0$ we divide all $g_i(z)$ with $(z-b)$.

Lemma 2.3. *The number of distinct zeros for $f(z)$ in (a,b) is given by*

$$\#\{i \mid g_{i-1}(a)g_i(a) < 0,\ 1 \le i \le m\} - \#\{i \mid g_{i-1}(b)g_i(b) < 0,\ 1 \le i \le m\}.$$

Another method which is not guaranteed to work, but is simpler when it does, is to rewrite $f(z)$ into the form

$$f(z) = \sum_{i=0}^{n} a_i (z-a)^i (b-z)^{n-i}$$

where $n = \deg(f(z))$. If $a_i \ge 0$ for all i, then clearly $f(z) \ge 0$ for all $z \in [a,b]$. For example, this is the method behind Theorem 2.5.

Example 2.7. As a further example, we show that the $[2^m, 2^m - m - 1, 4]$ extended Hamming code is proper. For convenience we write $M = 2^{m-1}$. Then

$$P_{\text{ue}}(p) = \frac{1}{4M}\Big\{1 + (4M-2)(1-2p)^M + (1-2p)^{2M}\Big\} - (1-p)^{2M},$$

and so

$$\begin{aligned}\frac{d}{dp}P_{\text{ue}}(p) &= 2M(1-p)^{2M-1} - (2M-1)(1-2p)^{M-1} - (1-2p)^{2M-1}\\ &= 2M(p+1-2p)^{2M-1}\\ &\quad -(2M-1)(2p+1-2p)^M(1-2p)^{M-1} - (1-2p)^{2M-1}\\ &= 2M\sum_{i=0}^{2M-1}\binom{2M-1}{i}p^i(1-2p)^{2M-1-i}\\ &\quad -(2M-1)\sum_{i=0}^{M}\binom{M}{i}2^ip^i(1-2p)^{2M-1-i} - (1-2p)^{2M-1}\\ &= \sum_{i=0}^{2M-1}\alpha_i p^i(1-2p)^{2M-1-i},\end{aligned}$$

where $\alpha_0 = \alpha_1 = \alpha_2 = 0$, and

$$i!\alpha_i = 2M(2M-1)(2M-2)\Big\{\prod_{j=3}^{i}(2M-j) - \prod_{j=3}^{i}(2M-2j+2)\Big\} > 0$$

for $3 \le i \le 2M-1$. Hence $\frac{d}{dp}P_{\text{ue}}(p) \ge 0$ for all $p \in \left[0, \frac{1}{2}\right]$; that is, C is proper.

2.3.2 *Sufficient conditions for a code to be good*

Theorem 2.5 may be useful to show that a code is good or proper. First,

$$P_{\text{ue}}\Big(C,\frac{q-1}{q}\Big)=\frac{M-1}{q^n}\Big(q\frac{p}{q-1}+1-\frac{q}{q-1}p\Big)^n$$
$$=\frac{M-1}{q^n}\sum_{i=0}^{n}\binom{n}{i}q^i\Big(\frac{p}{q-1}\Big)^i\Big(1-\frac{q}{q-1}p\Big)^{n-i}.$$

Therefore, if

$$A_i^\diamond\le\frac{q^i(M-1)}{q^n}\binom{n}{i}\qquad(2.5)$$

for $1\le i\le n$, then C is good. We can now use the results in Corollary 1.7. For $1\le i<d$ we have $A_i^\diamond=0$ and so (2.5) is satisfied. For $0\le n-i<d^\perp$ we have $A_i^\diamond=(Mq^{i-n}-1)\binom{n}{i}$ and so

$$\frac{q^i(M-1)}{q^n}\binom{n}{i}-A_i^\diamond=(1-q^{i-n})\binom{n}{i}\ge 0$$

and (2.5) is again satisfied. Therefore we get the following theorem.

Theorem 2.9. *Let C be an $(n,M,d;q)$ code with dual distance $d^\perp$. If*

$$A_i^\diamond\le\frac{(M-1)}{q^{n-i}}\binom{n}{i}$$

for $d\le i\le n-d^\perp$, then C is good.

2.3.3 *Necessary conditions for a code to be good or satisfactory*

We will give some necessary conditions for a code to be good.

Theorem 2.10. *Let C be a good $(n,M;q)$ code. Then, for all i, $0<i<n$ we have*

$$A_i(C)\le\frac{M-1}{q^n}\cdot\frac{n^n(q-1)^i}{i^i(n-i)^{n-i}}=\frac{M-1}{q^{n(1+H_q(i/n))}}.$$

Proof. Choosing $p=\frac{i}{n}$ we get

$$\frac{M-1}{q^n}\ge P_{\text{ue}}\Big(C,\frac{i}{n}\Big)=\sum_{j=1}^{n}A_j\Big(\frac{i}{n(q-1)}\Big)^j\Big(\frac{n-i}{n}\Big)^{n-j}$$
$$\ge A_i\Big(\frac{i}{n(q-1)}\Big)^i\Big(\frac{n-i}{n}\Big)^{n-i}=A_i\frac{i^i(n-i)^{n-i}}{n^n(q-1)^i}.$$

This proves the inequality. The equality follows directly from the definition of the q-ary entropy function $H_q(z)$ since

$$q^{-H_q(z)} = \Big(\frac{z}{q-1}\Big)^z (1-z)^{1-z}. \qquad \square$$

Theorem 2.11. *Let C be a satisfactory $(n, M; q)$ code. Then, for all i, $0 < i < n$ we have*

$$A_i(C) \le \frac{M}{q^n} \cdot \frac{n^n (q-1)^i}{i^i (n-i)^{n-i}},$$

$$A_i^\perp(C) \le \frac{1}{M} \cdot \frac{n^n (q-1)^i}{i^i (n-i)^{n-i}}.$$

Proof. The proof of the first inequality is similar to the proof of Theorem 2.10. For the second inequality, we first note that

$$\begin{aligned} Mq^{-n} &\ge P_{\text{ue}}(C,p) \\ &= Mq^{-n} + Mq^{-n} \sum_{j=1}^{n} A_j^\perp \Big(1 - \frac{q}{q-1}p\Big)^j - (1-p)^n \\ &\ge Mq^{-n} + Mq^{-n} A_i^\perp \Big(1 - \frac{q}{q-1}p\Big)^i - (1-p)^n \end{aligned}$$

and so

$$A_i^\perp \Big(1 - \frac{q}{q-1}p\Big)^i \le \frac{q^n}{M}(1-p)^n.$$

Choosing $1 - \frac{q}{q-1}p = \frac{i}{(n-i)(q-1)}$ we get

$$A_i^\perp \Big(\frac{i}{(n-i)(q-1)}\Big)^i \le \frac{1}{M}\Big(\frac{n}{n-i}\Big)^n. \qquad \square$$

Theorem 2.12. *Let C be a good $(n, M; q)$ code. Then*

$$MA_1^\perp \le (q-1)n.$$

Proof. From $P_{\text{ue}}(C,p) = Mq^{-n} \sum_{i=0}^{n} A_i^\perp \Big(1 - \frac{q}{q-1}p\Big)^i - (1-p)^n$ we get

$$\frac{d}{dp} P_{\text{ue}}(C,p) = n(1-p)^{n-1} - Mq^{-n} \sum_{i=1}^{n} \frac{qi}{q-1} A_i^\perp \Big(1 - \frac{q}{q-1}p\Big)^{i-1}.$$

Since $P_{\text{ue}}(C,p)$ cannot be decreasing at $p = \frac{q-1}{q}$ for a good code, we get

$$\begin{aligned} 0 &\le \frac{d}{dp} P_{\text{ue}}(C,p) \Big|_{p=\frac{q-1}{q}} \\ &= \frac{n}{q^{n-1}} - \frac{M}{q^n} \cdot \frac{q}{q-1} A_1^\perp \\ &= \frac{1}{(q-1)q^{n-1}} \Big\{(q-1)n - MA_1^\perp\Big\}. \end{aligned} \qquad \square$$

A class of necessary conditions can be obtained by considering a weighted average of $P_{\rm ue}(C,p)$ over some interval. We first formulate this as a general theorem, and thereafter we specialize it in two ways.

Theorem 2.13. *Let C be a good $(n,M;q)$ code. Let*

$$0 \le a < b \le (q-1)/q,$$

and let $f(p)$ be a continuous non-negative function on $[a,b]$ such that $\int_a^b f(p)dp = 1$ (that is, a probability distribution). Then

$$\int_a^b f(p)P_{\rm ue}(C,p)dp \le \frac{M-1}{q^n}.$$

Proof. Since C is good, $P_{\rm ue}(C,p) \le \frac{M-1}{q^n}$ for all $p \in [a,b]$, and so

$$\begin{aligned}\int_a^b f(p)P_{\rm ue}(C,p)dp &\le \int_a^b f(p)\frac{M-1}{q^n}dp \\ &= \frac{M-1}{q^n}\int_a^b f(p)dp = \frac{M-1}{q^n}.\end{aligned}$$

□

To compute the integral $\int_a^b f(p)P_{\rm ue}(C,p)dp$ we use one of the explicit expressions we have for $P_{\rm ue}(C,p)$. Let us first consider a constant $f(p)$, that is $f(p) = \frac{1}{b-a}$ for all p. For this case, it is most convenient to use Theorem 2.4, that is,

$$P_{\rm ue}(C,p) = \frac{M}{q^n}\sum_{i=0}^{n} A_i^{\perp}(C)\Big(1-\frac{qp}{q-1}\Big)^i - (1-p)^n.$$

This gives

$$\begin{aligned}\int_a^b f(p)P_{\rm ue}(C,p)dp &= \frac{M}{q^n(b-a)}\sum_{i=0}^{n} A_i^{\perp}(C)\int_a^b \Big(1-\frac{qp}{q-1}\Big)^i dp \\ &\quad -\frac{1}{(b-a)}\int_a^b (1-p)^n dp \\ &= \frac{M}{q^n(b-a)}\sum_{i=0}^{n} A_i^{\perp}(C)\frac{q-1}{q(i+1)} \\ &\qquad \cdot\Big\{\Big(1-\frac{qa}{q-1}\Big)^{i+1} - \Big(1-\frac{qb}{q-1}\Big)^{i+1}\Big\} \\ &\quad -\frac{1}{(b-a)(n+1)}\Big\{(1-a)^{n+1}-(1-b)^{n+1}\Big\}.\end{aligned}$$

This gives the following corollary.

Corollary 2.2. *Let C be a good $(n, M; q)$ code and let*

$$0 \le a < b \le (q-1)/q.$$

Then

$$\frac{M(q-1)}{q^{n+1}(b-a)} \sum_{i=0}^{n} \frac{A_i^{\perp}(C)}{i+1} \Big\{ \Big(1 - \frac{qa}{q-1}\Big)^{i+1} - \Big(1 - \frac{qb}{q-1}\Big)^{i+1} \Big\}$$
$$- \frac{1}{(b-a)(n+1)} \Big\{ (1-a)^{n+1} - (1-b)^{n+1} \Big\}$$
$$\le \frac{M-1}{q^n}.$$

For the next special case, we let $a = 0$ and $b = (q-1)/b$ and for $f(p)$ we use a so-called beta function. More precisely, let α and β be non-negative integers. Then (see e.g. Dweight (1961))

$$\int_0^1 x^\alpha (1-x)^\beta = \frac{\alpha!\beta!}{(\alpha+\beta)!} = \frac{1}{\binom{\alpha+\beta}{\alpha}}.$$

Hence

$$\int_0^{(q-1)/q} \Big(\frac{p}{q-1}\Big)^\alpha \Big(1 - \frac{qp}{q-1}\Big)^\beta dp = \frac{q-1}{(\alpha+\beta+1)q^{\alpha+1}\binom{\alpha+\beta}{\alpha}}.$$

Therefore, a possible choice for $f(p)$ is

$$f(p) = \frac{(\alpha+\beta+1)q^{\alpha+1}\binom{\alpha+\beta}{\alpha}}{q-1} \Big(\frac{p}{q-1}\Big)^\alpha \Big(1 - \frac{qp}{q-1}\Big)^\beta.$$

For the application of this weighting, the most useful form of $P_{\mathrm{ue}}(C,p)$ is the one given in Theorem 2.5:

$$P_{\mathrm{ue}}(C,p) = \sum_{i=1}^{n} A_i^{\diamond}(C) \Big(\frac{p}{q-1}\Big)^i \Big(1 - \frac{qp}{q-1}\Big)^{n-i}.$$

This gives

$$\int_0^{(q-1)/q} f(p) P_{\mathrm{ue}}(C,p) dp$$
$$= \frac{(\alpha+\beta+1)q^{\alpha+1}\binom{\alpha+\beta}{\alpha}}{q-1} \sum_{i=1}^{n} A_i^{\diamond}(C) \int_0^{(q-1)/q} \Big(\frac{p}{q-1}\Big)^{i+\alpha} \Big(1 - \frac{qp}{q-1}\Big)^{n-i+\beta} dp$$
$$= \frac{(\alpha+\beta+1)q^{\alpha+1}\binom{\alpha+\beta}{\alpha}}{q-1} \sum_{i=1}^{n} A_i^{\diamond}(C) \frac{q-1}{(n+\alpha+\beta+1)q^{i+\alpha+1}\binom{n+\alpha+\beta}{i+\alpha}}$$
$$= \frac{(\alpha+\beta+1)}{(n+\alpha+\beta+1)} \sum_{i=1}^{n} \frac{A_i^{\diamond}(C)}{q^i} \frac{\binom{\alpha+\beta}{\alpha}}{\binom{n+\alpha+\beta}{i+\alpha}}.$$

Since

$$\frac{(\alpha+\beta+1)\binom{\alpha+\beta}{\alpha}}{(n+\alpha+\beta+1)\binom{n+\alpha+\beta}{i+\alpha}} = \frac{\binom{\alpha+\beta+1}{\alpha+1}}{\binom{n+\alpha+\beta+1}{i+\alpha+1}},$$

we get the following corollary.

Corollary 2.3. *Let C be a good $(n, M; q)$ code and let α and β be non-negative integers. Then*

$$\sum_{i=1}^{n} \frac{A_i^{\diamond}(C)}{q^i} \frac{\binom{\alpha+\beta+1}{\alpha+1}}{\binom{n+\alpha+\beta+1}{i+\alpha+1}} \leq \frac{M-1}{q^n}.$$

For a linear $[n, k, d; q]$ code, a general lower bound on A_d is $q-1$, and for a non-linear $(n, M, d; q)$ code a general lower bound on A_d is $2/M$. Now, let A be some positive number. We will consider $(n, M, d; q)$ codes for which $A_d \geq A$. In the rest of the subsection we also use the notations

$$Q = \frac{q}{q-1} \text{ and } \kappa = \ln(M/A) = \ln M - \ln A.$$

By definition,

$$P_{\mathrm{ue}}^{\perp}(C, p) \geq \frac{1}{M} + \frac{A}{M}(1-Qp)^d - (1-p)^n.$$

Hence, if

$$\frac{A}{M}(1-Qp)^d \geq (1-p)^n, \tag{2.6}$$

then $P_{\mathrm{ue}}^{\perp}(C, p) \geq \frac{1}{M}$. Taking logarithms in (2.6), we get the equivalent condition

$$-\kappa + d\ln(1-Qp) \geq n\ln(1-p).$$

Combining this with Theorem 2.8, 1), we get the following lemma.

Lemma 2.4. *If C is an $(n, M, d)_q$ code and*

$$n \geq h(p) = \frac{d\ln(1-Qp) - \kappa}{\ln(1-p)},$$

then C is ugly.

Any choice of p, $0 < p < (q-1)/q$ now gives a proof of the existence of a $\mu(d, \kappa)$ such that if $n \geq \mu(d, \kappa)$ and C is an (n, M, d) code with $A_d \geq A$, then C is ugly for error detection. The strongest result is obtained for the p that minimizes $h(p)$. We cannot find a closed formula for this, but consider approximations.

We will use the notations

$$f(p) = \frac{\ln(1-Qp)}{\ln(1-p)}, \text{ and } g(p) = \frac{-1}{\ln(1-p)}.$$

Then

$$h(p) = d\,f(p) + \kappa\, g(p). \tag{2.7}$$

The function $f(p)$ is increasing on $(0,(q-1)/q)$, it approaches the value Q when $p \to 0+$, and it approaches infinity when $p \to (q-1)/q-$. Moreover,

$$f'(p) = \frac{-Q(1-p)\ln(1-p) + (1-Qp)\ln(1-Qp)}{(1-p)(1-Qp)\ln(1-p)^2},$$

and

$$f''(p) = \frac{f_1(p)}{-(1-p)^2(1-Qp)^2(\ln(1-p))^3},$$

where

$$\begin{aligned} f_1(p) &= Q^2(1-p)^2(\ln(1-p))^2 + 2Q(1-p)(1-Qp)\ln(1-p) \\ &\quad -2(1-Qp)^2\ln(1-Qp) - (1-Qp)^2\ln(1-p)\ln(1-Qp) \\ &> 0 \end{aligned}$$

for all $p \in (0,(q-1)/q)$. Hence f is convex on $(0,(q-1)/q)$. Similarly, the function $g(p)$ is decreasing on $(0,(q-1)/q)$, it approaches infinity when $p \to 0+$, and it takes the value $-1/\ln q$ for $p = (q-1)/q$. Moreover,

$$g'(p) = \frac{-1}{(1-p)\ln(1-p)^2} = \frac{-(1-Qp)}{(1-p)(1-Qp)\ln(1-p)^2},$$

$$g''(p) = \frac{-(2+\ln(1-p))}{(1-p)^2(\ln(1-p))^3} > 0$$

for all $p \in (0,(q-1)/q)$, and so $g(p)$ is also convex on $(0,(q-1)/q)$. This implies that the combined function $h(p)$ is also convex on $(0,(q-1)/q)$ since $\kappa > 0$, and it takes its minimum somewhere in $(0,(q-1)/q)$. We denote the value of p where minimum is obtained by p_m and the minimum by $\mu(d,\kappa)$.

From Lemma 2.4 we get the following necessary condition for a code to be good.

Corollary 2.4. *If C is good for error detection, then $n < \mu(d,\kappa)$.*

One can find good approximations for $\mu(d,\kappa)$ when $d \gg k$ or $k \gg d$. Here we give some details for $d \geq k$. Let $\kappa = \alpha d$, where α is a parameter, $0 \leq \alpha \leq 1$. Then

$$h(p) = d\,\frac{\ln(1-Qp) - \alpha}{\ln(1-p)}$$

and

$$\frac{h'(p)}{d} = \frac{-Q(1-p)\ln(1-p) + (1-Qp)\ln(1-Qp)}{(1-p)(1-Qp)\ln(1-p)^2} - \frac{\alpha}{(1-p)\ln(1-p)^2}.$$

In particular $h'(p) = 0$ if (and only if)

$$\alpha = \frac{-Q(1-p)\ln(1-p) + (1-Qp)\ln(1-Qp)}{1-Qp}. \tag{2.8}$$

We want to solve this for p in terms of α. There is no closed form of this solution. However, we can find good approximations. For $\alpha \to 0+$, we see that $p \to 0$ and $h(p) \to Q$. We will first describe this important case in more detail. We note that $\alpha \to 0+$ implies that $d \to \infty$. The parameter κ may also grow, but then at a slower rate (since $d/\kappa \to 0$).

Theorem 2.14. *Let*

$$y = \sqrt{\frac{\alpha}{2Q(Q-1)}}.$$

There exist numbers a_i and b_i for $i = 1, 2, \ldots$ such that, for any $r \geq 0$,

$$p_m = \sum_{i=1}^{r} a_i y^i + O(y^{r+1}),$$

and

$$\mu(d,\alpha d) = dQ\left\{1 + 2(Q-1)\sum_{i=1}^{r} b_i y^i + O(y^{r+1})\right\}$$

when $y \to 0$ (that is $\alpha \to 0$). The first few a_i and b_i are given in Table 2.6.

Proof. First we note that $\alpha = 2Q(Q-1)y^2$ and so

$$h(p) = d\,\frac{\ln(1-Qp) - 2Q(Q-1)y^2}{\ln(1-p)},$$

and

$$h'(p) = d\,\frac{H(p,y)}{(1-p)(1-Qp)(\ln(1-p))^2},$$

Table 2.6 a_i and b_i for Theorem 2.14

i	a_i	b_i
1	2	1
2	$-(8Q+2)/3$	$(2Q-1)/3$
3	$(26Q^2+22Q-1)/9$	$(2Q^2-2Q-1)/18$
4	$-(368Q^3+708Q^2-12Q+8)/135$	$-2(Q-2)(2Q-1)(Q+1)/135$

where

$$H(p,y) = -Q(1-p)\ln(1-p) + (1-Qp)\ln(1-Qp) - 2Q(Q-1)y^2(1-Qp).$$

Hence $h'(p) = 0$ if $H(p,y) = 0$. Taking the Taylor expansion of $H(\sum a_i y^i, y)$ we get

$$H\left(\sum a_i y^i, y\right) = \frac{a_1^2-4}{4}y^2 + \frac{a_1}{6}(Qa_1^2 + a_1^2 + 6a_2 + 12Q)y^3 + \cdots$$

All coefficients for $i \le r$ should be zero. In particular, the coefficient of y^2 shows that $a_1^2 = 4$. Since $a_1 y^2$ is the dominating term in the expression for p when y is small and $p > 0$, we must have $a_1 > 0$ and so $a_1 = 2$. Next the coefficient of y^3 shows that $a_2 = -(16Q+4)/6$. In general, we get equations in the a_i which can be used to determine the a_i recursively. Substituting the expression for p into $h(p)$ and taking Taylor expansion, we get the expression for $\mu(d,\kappa)$. □

Assuming that $\kappa\alpha \to 0$ and taking the first three terms of approximation, we get

$$\mu(d,\kappa) \approx dQ + \sqrt{2d\kappa Q(Q-1)} + \frac{2Q-1}{3}\kappa,$$

(the other terms goes to zero with y).

By definition, $h(p)$ is an upper approximation for any p. One way to get a good upper approximation is to choose for p a good approximation for p_m. For example, taking the first term in the approximation for p_m, that is, $p = \sqrt{2\alpha/(Q(Q-1))}$, we get

$$\mu(d,\kappa) \le h\left(\sqrt{2\alpha/(Q(Q-1))}\right).$$

By similar analysis, one can determine approximations for $\mu(d,\kappa)$ when κ is larger than d. The main term is

$$\mu(d,\kappa) \approx \frac{\kappa}{\ln q}.$$

2.3.4 *Sufficient conditions for a code to be proper*

An immediate consequence of Theorem 2.2 is the following theorem.

Theorem 2.15. *If C is an $(n, M, d; q)$ code and $d \ge \frac{q-1}{q}n$, then C is proper.*

A related condition is the following.

Theorem 2.16. *If C is an $(n, M, d; q)$ code and $d^\perp > \frac{q-1}{q}n$, then C is proper.*

Proof. Any code of length one is proper, so we may assume that $n \ge 2$. Let

$$f(p) = P_{\rm ue}(C,p) = \frac{M}{q^n} + \frac{M}{q^n}\sum_{i=d^\perp}^{n}\Big(1-\frac{qp}{q-1}\Big)^i - (1-p)^n.$$

Then

$$f'(p) = n(1-p)^{n-1} - \frac{M}{q^n}\sum_{i=d^\perp}^{n} iA_i^\perp \frac{q}{q-1}\Big(1-\frac{qp}{q-1}\Big)^{i-1}.$$

Hence

$$\frac{f'(p)}{n(1-p)^{n-1}} = 1 - \frac{M}{n(q-1)q^{n-1}}\sum_{i=d^\perp}^{n} iA_i^\perp \frac{\Big(1-\frac{qp}{q-1}\Big)^{i-1}}{(1-p)^{n-1}}. \tag{2.9}$$

We note that $d^\perp > \frac{q-1}{q}n$ implies that

$$d^\perp - 1 \ge \frac{(q-1)n+1}{q} - 1 = \frac{q-1}{q}(n-1).$$

Hence, for all $i \ge d^\perp$,

$$\Big(1-\frac{qp}{q-1}\Big)^{i-1} \le \Big(1-\frac{qp}{q-1}\Big)^{d^\perp-1} \le \Big(1-\frac{qp}{q-1}\Big)^{(n-1)(q-1)/q}$$

and so

$$\frac{\Big(1-\frac{qp}{q-1}\Big)^{i-1}}{(1-p)^{n-1}} \le \left(\frac{1-\frac{q}{q-1}p}{(1-p)^{q/(q-1)}}\right)^{(q-1)(n-1)/q} \le 1.$$

Therefore, by (1.4)

$$\begin{aligned}\frac{f'(p)}{n(1-p)^{n-1}} &\ge 1 - \frac{M}{n(q-1)q^{n-1}}\sum_{i=d^\perp}^{n} iA_i^\perp \\ &= 1 - \frac{M}{n(q-1)q^{n-1}}\cdot\frac{q^{n-1}}{M}\{(q-1)n - A_1\} \ge 0.\end{aligned}$$

Hence, $f'(p) \ge 0$ for all $p \in [0, (q-1)/q]$ and so C is proper. □

Remark 2.1. The bound in the theorem cannot be improved in general, in the sense that there exist bad $(n, M, d; q)$ codes for which $d^\perp = \frac{q-1}{q}n$.

The idea of the proof of Theorem 2.16 can be carried further. Let $0 < \alpha < (q-1)/q$. If $h(p) = \left(1 - \frac{q}{q-1}p\right)^\alpha$, then

$$h'(p) = -\alpha\frac{q}{q-1}\left(1 - \frac{q}{q-1}p\right)^{\alpha-1} < 0$$

and

$$h''(p) = \alpha(\alpha-1)\left(\frac{q}{q-1}\right)^2\left(1 - \frac{q}{q-1}p\right)^{\alpha-2} < 0.$$

Hence, the equation

$$\left(1 - \frac{q}{q-1}p\right)^\alpha = 1 - p$$

has a unique solution $\rho(q, \alpha)$ in the interval $0 < p < \frac{q-1}{q}$. We observe that if $\alpha < \beta$, then $\left(1 - \frac{q}{q-1}p\right)^\alpha < \left(1 - \frac{q}{q-1}p\right)^\beta$ for all p. Hence, $\rho(q, \alpha) > \rho(q, \beta)$.

Theorem 2.17. *If C is an $(n, M, d; q)$ code and*

$$\frac{d}{n} \geq \rho\left(q, \frac{d^\perp - 1}{n-1}\right),$$

then C is proper.

Proof. By (2.9),

$$\begin{aligned}
\frac{f'(p)}{n(1-p)^{n-1}} &\geq 1 - \frac{\left(1 - \frac{qp}{q-1}\right)^{d^\perp - 1}}{(1-p)^{n-1}} \frac{M}{n(q-1)q^{n-1}} \sum_{i=d^\perp}^{n} iA_i^\perp \\
&= 1 - \frac{\left(1 - \frac{qp}{q-1}\right)^{d^\perp - 1}}{(1-p)^{n-1}} \frac{M}{n(q-1)q^{n-1}} \cdot \frac{q^{n-1}}{M}\{(q-1)n - A_1\} \\
&\geq 1 - \left\{\frac{\left(1 - \frac{qp}{q-1}\right)^{(d^\perp-1)/(n-1)}}{(1-p)}\right\}^{n-1} \geq 0
\end{aligned}$$

for

$$\rho\left(q, \frac{d^\perp - 1}{n-1}\right) \leq p \leq \frac{q-1}{q}.$$

Hence, $f(p)$ is increasing in this interval. On the other hand, by Theorem 2.2, $f(p)$ is increasing on $[0, d/n]$. Combining these two results, the theorem follows. □

Example 2.8. The equation for $\rho(2,1/3)$ is $(1-2p)^{1/3} = 1-p$, that is, $1-2p = (1-p)^3$ which has the solution $\rho(2,1/3) = (3-\sqrt{5})/2 \approx 0.3819$. Hence, if C is an $(n, M, d; 2)$ code such that $(d^\perp - 1)/(n-1) \geq 1/3$ and $d/n \geq 0.3819$, then C is proper.

Example 2.9. Consider a code C for which $d \geq 2$ and $d^\perp = \frac{q-1}{q}n$. Let $\alpha = \left(\frac{q-1}{q}n - 1\right)/(n-1)$. If

$$\left(1 - \frac{q}{q-1}\frac{2}{n}\right)^\alpha \leq 1 - \frac{2}{n}, \tag{2.10}$$

then $\rho(q,\alpha) \leq 2/n$ and the code C is proper by Theorem 2.17. A careful analysis shows that (2.10) is satisfied for $n \geq 5$ when $q = 2$ and for $n \geq 3$ when $q \geq 3$.

To formulate the next sufficient condition for a code being proper, we define the functions

$$\Lambda_i(p) = \sum_{j=i}^{n} \binom{n}{j} \left(\frac{qp}{q-1}\right)^j \left(1 - \frac{qp}{q-1}\right)^{n-j}.$$

Theorem 2.18. *Let C be an $(n, M; q)$ code. Then*

$$P_{\text{ue}}(C,p) = \frac{A_1^\diamond(C)}{\binom{n}{i}q}\Lambda_1(p) + \sum_{i=2}^{n}\left\{\frac{A_i^\diamond(C)}{\binom{n}{i}q^i} - \frac{A_{i-1}^\diamond(C)}{\binom{n}{i-1}q^{i-1}}\right\}\Lambda_i(p).$$

Proof. Since

$$\Lambda_i(p) - \Lambda_{i+1}(p) = \binom{n}{i} q^i \left(\frac{p}{q-1}\right)^i \left(1 - \frac{qp}{q-1}\right)^{n-i},$$

Theorem 2.5 implies that

$$\begin{aligned} P_{\text{ue}}(C,p) &= \sum_{i=1}^{n} A_i^\diamond(C) \frac{1}{\binom{n}{i}q^i}(\Lambda_i(p) - \Lambda_{i+1}(p)) \\ &= \sum_{i=1}^{n} \frac{A_i^\diamond(C)}{\binom{n}{i}q^i}\Lambda_i(p) - \sum_{i=2}^{n} \frac{A_{i-1}^\diamond(C)}{\binom{n}{i-1}q^{i-1}}\Lambda_i(p) \\ &= \frac{A_1^\diamond(C)}{\binom{n}{i}q}\Lambda_1(p) + \sum_{i=2}^{n}\left\{\frac{A_i^\diamond(C)}{\binom{n}{i}q^i} - \frac{A_{i-1}^\diamond(C)}{\binom{n}{i-1}q^{i-1}}\right\}\Lambda_i(p). \end{aligned}$$

□

The representation in Theorem 2.18 is useful because of the following result.

Lemma 2.5. *The functions $\Lambda_i(p)$ are increasing on $[0, \frac{q-1}{q}]$ for all $i \geq 1$.*

Proof. This is shown by straightforward calculus:

$$\begin{aligned}
\frac{d\Lambda_i(p)}{dp} &= \sum_{j=i}^{n} \binom{n}{j} \Big\{ j\Big(\frac{qp}{q-1}\Big)^{j-1} \frac{q}{q-1}\Big(1-\frac{qp}{q-1}\Big)^{n-j} \\
&\quad -(n-j)\Big(\frac{qp}{q-1}\Big)^{j}\Big(1-\frac{qp}{q-1}\Big)^{n-j-1}\frac{q}{q-1}\Big\} \\
&= \frac{q}{q-1}\sum_{j=i-1}^{n}\binom{n}{j+1}(j+1)\Big(\frac{qp}{q-1}\Big)^{j}\Big(1-\frac{qp}{q-1}\Big)^{n-j-1} \\
&\quad -\frac{q}{q-1}\sum_{j=i}^{n}\binom{n}{j}(n-j)\Big(\frac{qp}{q-1}\Big)^{j}\Big(1-\frac{qp}{q-1}\Big)^{n-j-1} \\
&= \frac{q}{q-1}\binom{n}{i} i\Big(\frac{qp}{q-1}\Big)^{i-1}\Big(1-\frac{qp}{q-1}\Big)^{n-i} > 0
\end{aligned}$$

for all $p \in (0, \frac{q-1}{q})$. □

Combining Theorem 2.18, Corollary 1.7, and Lemma 2.5, we get the following result.

Theorem 2.19. *Let C be an $(n, M; q)$ code. If*

$$\frac{A_i^{\diamond}(C)}{\binom{n}{i}} \geq q\frac{A_{i-1}^{\diamond}(C)}{\binom{n}{i-1}} \tag{2.11}$$

for $d+1 \leq i \leq n-d^{\perp}$, then C is proper.

Remark 2.2. The condition (2.11) can be rewritten as

$$iA_i^{\diamond}(C) \geq q(n-i+1)A_{i-1}^{\diamond}(C) \tag{2.12}$$

and also

$$\sum_{j=1}^{i} \frac{i_{(j)}}{n_{(j)}} A_j(C) \geq q \sum_{j=1}^{i-1} \frac{(i-1)_{(j)}}{n_{(j)}} A_j(C), \tag{2.13}$$

where

$$m_{(j)} = m(m-1)...(m-j+1).$$

2.3.5 *Large codes are proper*

The code F_q^n is proper: $P_{\text{ue}}(F_q^n, p) = 1-(1-p)^n$ and this is clearly increasing with p. Hence, there exists a bound $B(q,n) \leq q^n$ such that if

$M \geq B(q,n)$, then *any* (n, M, q) code is proper. We will give estimates for $B(q,n)$. Let

$$\beta(q,n) = \frac{-(qn - n + q) + s}{2(qn - n - q)},$$

where

$$s = \sqrt{(qn - n + q)^2 + 4n(q-1)(qn - n - q)q^n}.$$

Lemma 2.6. *If* $|C| \leq \beta(q,n)$, *then* $\overline{C}$ *is proper.*

We note that if $|\overline{C}| \geq q^n - \beta(q,n)$, then

$$|C| \leq \beta(q,n).$$

Hence from the lemma we get the following equivalent result.

Theorem 2.20. *For* $q \geq 2$ *and* $n \geq 3$ *we have*

$$B(q,n) \leq q^n - \beta(q,n);$$

that is, any q-ary code of size at least $q^n - \beta(q,n)$ *is proper.*

Proof. Combining Corollary 1.3 and Theorem 2.19 (with the condition in the form (2.13)), a sufficient condition for $\overline{C}$ to be proper is

$$\sum_{i=1}^{l} \frac{l_{(i)}}{n_{(i)}} \{MA_i + (q^n - 2M)\binom{n}{i}(q-1)^i\}$$
$$\geq q \sum_{i=1}^{l-1} \frac{(l-1)_{(i)}}{n_{(i)}} \{MA_i + (q^n - 2M)\binom{n}{i}(q-1)^i\}.$$

We also note that

$$\sum_{i=1}^{l} \frac{l_{(i)}}{n_{(i)}} \binom{n}{i}(q-1)^i = \sum_{i=1}^{l} \binom{l}{i}(q-1)^i = q^l - 1$$

for $2 \leq l \leq n$. Hence, the sufficient condition for $\overline{C}$ to be proper can be written

$$M \sum_{i=1}^{l} \frac{l_{(i)}}{n_{(i)}} A_i + (q^n - 2M)(q^l - 1)$$
$$\geq qM \sum_{i=1}^{l-1} \frac{(l-1)_{(i)}}{n_{(i)}} A_i + (q^n - 2M)q(q^{l-1} - 1) \tag{2.14}$$

for $2 \leq l \leq n$. Omitting the term for $i = l$ in the sum on the left-hand side and rearranging, we get the following stronger sufficient condition for

$\overline{C}$ to be proper (stronger in the sense that if (2.15) is satisfied, then (2.14) is also satisfied):

$$(q^n - 2M)(q-1) \geq M \sum_{i=1}^{l-1} \frac{(l-1)_{(i)}}{n_{(i)}} \left(q - \frac{l}{l-i}\right) A_i \tag{2.15}$$

for $2 \leq l \leq n$. Since

$$\frac{(l-1)_{(i)}}{n_{(i)}} \left(q - \frac{l}{l-i}\right) = \frac{1}{n} \frac{(l-1)_{(i-1)}}{(n-1)_{(i-1)}} (ql - qi - l) \leq \frac{1}{n}(ql - q - l),$$

we get

$$\sum_{i=1}^{l-1} \frac{(l-1)_{(i)}}{n_{(i)}} \left(q - \frac{l}{l-i}\right) A_i \leq \frac{1}{n}(ql - l - q) \sum_{i=1}^{l-1} A_i \leq \frac{1}{n}(ql - l - q)(M-1). \tag{2.16}$$

Combining (2.16) with (2.15), we see that a yet stronger sufficient condition for $\overline{C}$ to be proper is

$$(q^n - 2M)(q-1) \geq \frac{1}{n}(ql - l - q)M(M-1) \tag{2.17}$$

for $2 \leq l \leq n$. The strongest condition is imposed for $l = n$ and this condition is

$$(q^n - 2M)(q-1) \geq \frac{1}{n}(qn - n - q)M(M-1).$$

This is equivalent to

$$M \leq \beta(q, n).$$

This completes the proof of Lemma 2.6 and Theorem 2.20. □

Using different estimates, we can obtain a stronger bound than Theorem 2.20; this bound is not explicit, however, but needs some computation. In (2.15), the terms in the sum are non-positive for $q - l/(l-i) < 0$, that is $i > l(q-1)/q$. If we omit these terms, we get the stronger conditions

$$(q^n - 2M)(q-1) \geq M \sum_{i=1}^{\lfloor l(q-1)/q \rfloor} \frac{(l-1)_{(i)}}{n_{(i)}} \left(q - \frac{l}{l-i}\right) A_i \tag{2.18}$$

for $2 \leq l \leq n$. We note that the condition for $l+1$ is stronger than the condition for l since

$$\frac{(l)_{(i)} \left(q - \frac{l+1}{l+1-i}\right)}{(l-1)_{(i)} \left(q - \frac{l}{l-i}\right)} = \frac{l(ql + q - qi - l - 1)}{(l+1-i)(ql - qi - l)} \geq 1$$

for $i \le l(q-1)/q$. Hence, the strongest condition is the condition for $l = n$, namely

$$(q^n - 2M)(q-1) \ge M \sum_{i=1}^{\lfloor n(q-1)/q \rfloor} \frac{1}{n}(q(n-i)-n)A_i. \tag{2.19}$$

We have $A_i \le \binom{n}{i}(q-1)^i$. We can use this bound on A_i, but must take into account that $\sum_{i=1}^{n} A_i = M - 1$. Let

$$M_m = \sum_{i=1}^{m} \binom{n}{i}(q-1)^i.$$

Assume that $M > M_{r-1}$ where $r < \lfloor n(q-1)/q \rfloor$, and let

$$S = \sum_{i=1}^{r-1}(q(n-i)-n)\binom{n}{i}(q-1)^i + (q(n-r)-n)(M-1-M_{r-1}).$$

Then

$$\begin{aligned}
S &\ge \sum_{i=1}^{r-1}(q(n-i)-n)\binom{n}{i}(q-1)^i + (q(n-r)-n)\sum_{i=1}^{\lfloor n(q-1)/q \rfloor} A_i \\
&\quad -(q(n-r)-n)\sum_{i=1}^{r-1}\binom{n}{i}(q-1)^i \\
&= \sum_{i=1}^{r-1} q(r-i)\binom{n}{i}(q-1)^i + (q(n-r)-n)\sum_{i=1}^{\lfloor n(q-1)/q \rfloor} A_i \\
&= \sum_{i=1}^{\lfloor n(q-1)/q \rfloor} ((q-i)-n)A_i + \sum_{i=1}^{r-1} q(r-i)\Big(\binom{n}{i}(q-1)^i - A_i\Big) \\
&\quad + \sum_{i=r+1}^{\lfloor n(q-1)/q \rfloor} q(i-r)A_i \\
&\ge \sum_{i=1}^{\lfloor n(q-1)/q \rfloor} ((q-i)-n)A_i.
\end{aligned}$$

Hence, by (2.19) we get the following stronger sufficient condition for $\overline{C}$ to be proper:

$$\begin{aligned}
(q^n - 2M)(q-1) \ge \frac{M}{n}&\sum_{i=1}^{r-1}(q(n-i)-n)\binom{n}{i}(q-1)^i \\
&+\frac{M}{n}(q(n-r)-n)(M-1-M_{r-1}).
\end{aligned}$$

Solving this inequality, we get a maximal value which we denote by $\beta_r(q,n)$. Provided $\beta_r(q,n) > M_{r-1}$ (this was a requirement for the derivation above) we have

$$B(n,q) \le q^n - \beta_r(q,n).$$

In particular, $\beta_1(q,n) = \beta(q,n)$. The larger r is, the larger is the corresponding $\beta_r(q,n)$. Therefore, we want to choose r as large as possible. We know that we can use r determined by $M_{r-1} < \beta(q,n) \le M_r$. Usually, but not always, this is maximal.

Example 2.10. We illustrate the procedure by a numerical example, namely $q = 3$ and $n = 21$. The values of M_r and $\beta_r(3,21)$ are given in Table 2.7. Note that this is an example where r determined by $M_{r-1} < \beta(q,n) \le M_r$ is $r = 4$, whereas the maximal r that can be used is larger, namely, $r = 5$.

Table 2.7 Table for Example 2.10

r	M_{r-1}	$\beta_r(3,21)$
1	0	106136.11
2	42	110468.15
3	882	115339.98
4	11522	120392.84
5	107282	121081.17
6	758450	

A lower bound on $B(q,n)$ is obtained by giving an explicit code which is not proper. Let

$$\gamma(2,n) = 2^{\lfloor (n+3)/2 \rfloor}$$
$$\gamma(q,n) = q^{\lfloor (n+2)/2 \rfloor} \text{ for } q \ge 3.$$

Theorem 2.21. *For $q \ge 2$ and $n \ge 4$ we have*

$$B(q,n) > q^n - \gamma(q,n).$$

Proof. Let C be the $(n, q^k; q)$ code

$$\{(\mathbf{x}|\mathbf{0}) \in F_q^n \mid \mathbf{x} \in F_q^k\}.$$

It is easy to see that

$$A_i = \binom{k}{i}(q-1)^i$$

for all i. Hence,

$$P_{\mathrm{ue}}(C,p) = \sum_{i=1}^{k} \binom{k}{i} p^i (1-p)^{k-i} (1-p)^{n-k} = (1-p)^{n-k} - (1-p)^n.$$

Therefore, if $f(p) = (q^n - q^k) P_{\mathrm{ue}}(\overline{C}, p)$, Corollary 1.3 shows that

$$\begin{aligned} f(p) &= q^k((1-p)^{n-k} - (1-p)^n) + (q^n - 2q^k)(1-(1-p)^n) \\ &= q^n - 2q^k + q^k(1-p)^{n-k} - (q^n - q^k)(1-p)^n. \end{aligned}$$

Hence,

$$f'(p) = -(n-k)q^k(1-p)^{n-k-1} + n(q^n - q^k)(1-p)^{n-1},$$

and so

$$f'\Big(\frac{q-1}{q}\Big) = -(n-k)q^{2k+1-n} + n(q^n - q^k)q^{-n+1} < 0$$

if

$$-q^k(n-k) + n(q^{n-k} - 1) < 0. \tag{2.20}$$

Let $k = (n+\alpha)/2$ (where $\alpha = 1, 2, 3$). Then (2.20) is equivalent to

$$q^{(n-\alpha)/2}\Big(n - q^{\alpha}\frac{n-\alpha}{2}\Big) - n < 0.$$

For $\alpha = 3$ and $q \geq 2$ we have

$$n - q^{\alpha}\frac{n-\alpha}{2} \leq n - 2^3\frac{n-3}{2} \leq 0 \text{ for } n \geq 4.$$

For $\alpha = 2$ and $q \geq 2$ we have

$$n - q^{\alpha}\frac{n-\alpha}{2} \leq n - 2^2\frac{n-2}{2} \leq 0 \text{ for } n \geq 4.$$

For $\alpha = 1$ and $q \geq 3$ we have

$$n - q^{\alpha}\frac{n-\alpha}{2} \leq n - 3\frac{n-1}{2} \leq 0 \text{ for } n \geq 3.$$

Hence, $f'\big(\frac{q-1}{q}\big) < 0$ for $q = 2$, $k = \lfloor (n+3)/2 \rfloor$, and $n \geq 4$; and also for $q \geq 3$, $k = \lfloor (n+2)/2 \rfloor$, and $n \geq 4$. □

2.4 Results on the average probability

2.4.1 *General results on the average*

In this section we consider the average probability of undetected error for the codes in some set $\mathcal{C}$ of codes of length n.

The common notations for the *average* and *variance* of a stochastic variable X are $E(X)$ and $Var(X) = E(X^2) - E(X)^2$. Further $\sigma = \sqrt{Var(X)}$, the *standard deviation* of X.

For our particular application use the notations

$$P_{\text{ue}}(\mathcal{C}, p) = \frac{1}{\#\mathcal{C}} \sum_{C \in \mathcal{C}} P_{\text{ue}}(C, p),$$

and

$$Var(\mathcal{C}, p) = Var(\{P_{\text{ue}}(C, p) \mid C \in \mathcal{C}\}).$$

For $S \subseteq F_q^n$, let

$$\alpha(S) = \alpha_{\mathcal{C}}(S) = \frac{\#\{C \in \mathcal{C} \mid S \subseteq C\}}{\#\mathcal{C}}.$$

If $S = \{\mathbf{x}_1, \mathbf{x}_2, \ldots \mathbf{x}_r\}$, we write for convenience $\alpha(S) = \alpha(\mathbf{x}_1, \mathbf{x}_2, \ldots \mathbf{x}_r)$.

From the definitions and Theorem 2.1 we get the following result.

Theorem 2.22. *Let $\mathcal{C}$ be a set of codes of length n and size M. Then*

$$P_{\text{ue}}(\mathcal{C}, p) = \frac{1}{M} \sum_{\substack{(\mathbf{x},\mathbf{y}) \in (Z_q^n)^2 \\ \mathbf{x} \neq \mathbf{y}}} \alpha(\mathbf{x}, \mathbf{y}) \Big(\frac{p}{q-1}\Big)^{d_H(\mathbf{x},\mathbf{y})} (1-p)^{n-d_H(\mathbf{x},\mathbf{y})}.$$

Proof. We have

$$\begin{aligned}
P_{\text{ue}}(\mathcal{C}, p) &= \frac{1}{\#\mathcal{C}} \sum_{C \in \mathcal{C}} \frac{1}{M} \sum_{\substack{(\mathbf{x},\mathbf{y}) \in C^2 \\ \mathbf{x} \neq \mathbf{y}}} \Big(\frac{p}{q-1}\Big)^{d_H(\mathbf{x},\mathbf{y})} (1-p)^{n-d_H(\mathbf{x},\mathbf{y})} \\
&= \frac{1}{M} \sum_{\substack{(\mathbf{x},\mathbf{y}) \in (Z_q^n)^2 \\ \mathbf{x} \neq \mathbf{y}}} \Big(\frac{p}{q-1}\Big)^{d_H(\mathbf{x},\mathbf{y})} (1-p)^{n-d_H(\mathbf{x},\mathbf{y})} \frac{1}{\#\mathcal{C}} \sum_{\substack{C \in \mathcal{C} \\ \mathbf{x},\mathbf{y} \in C}} 1 \\
&= \frac{1}{M} \sum_{\substack{(\mathbf{x},\mathbf{y}) \in (Z_q^n)^2 \\ \mathbf{x} \neq \mathbf{y}}} \alpha(\mathbf{x}, \mathbf{y}) \Big(\frac{p}{q-1}\Big)^{d_H(\mathbf{x},\mathbf{y})} (1-p)^{n-d_H(\mathbf{x},\mathbf{y})}.
\end{aligned}$$

□

For linear codes we get a simpler expression, using Theorem 1.13.

Theorem 2.23. *Let $\mathcal{C}$ be a set of linear codes of length n. Then*

$$P_{\text{ue}}(\mathcal{C}, p) = \sum_{\substack{\mathbf{x} \in GF(q)^n \\ \mathbf{x} \neq \mathbf{0}}} \alpha(\mathbf{x}) \Big(\frac{p}{q-1}\Big)^{w_{\text{H}}(\mathbf{x})} (1-p)^{n - w_{\text{H}}(\mathbf{x})}.$$

Let

$$A_i(\mathcal{C}) = \frac{1}{\#\mathcal{C}} \sum_{C \in \mathcal{C}} A_i(C).$$

Then clearly

$$P_{\text{ue}}(\mathcal{C}, p) = \sum_{i=1}^{n} A_i(\mathcal{C}) \Big(\frac{p}{q-1}\Big)^{i} (1-p)^{n-i}.$$

From Theorems 2.22 and 2.23 we get the following corollaries.

Corollary 2.5. *Let $\mathcal{C}$ be a set of all codes of length n and size M. Then*

$$A_i(\mathcal{C}) = \frac{1}{M} \sum_{\substack{(\mathbf{x},\mathbf{y}) \in (F_q^n)^2 \\ d_{\text{H}}(\mathbf{x},\mathbf{y}) = i}} \alpha(\mathbf{x}, \mathbf{y}).$$

Corollary 2.6. *Let $\mathcal{C}$ be a set of all linear codes of length n and dimension k. Then*

$$A_i(\mathcal{C}) = \sum_{\substack{\mathbf{x} \in GF(q)^n \\ w_{\text{H}}(\mathbf{x}) = i}} \alpha(\mathbf{x}).$$

2.4.2 *The variance*

For the variance we get similar results.

Theorem 2.24. *Let $\mathcal{C}$ be a set of codes of length n and size M. Then*

$$Var(\mathcal{C}, p) = -P_{\text{ue}}(\mathcal{C}, p)^2 + \frac{1}{M^2} \sum_{\substack{(\mathbf{u},\mathbf{v}),(\mathbf{x},\mathbf{y}) \in (F_q^n)^2 \\ \mathbf{u} \neq \mathbf{v}, \mathbf{x} \neq \mathbf{y}}} \alpha(\mathbf{u}, \mathbf{v}, \mathbf{x}, \mathbf{y})$$

$$\cdot \Big(\frac{p}{q-1}\Big)^{d_{\text{H}}(\mathbf{u},\mathbf{v}) + d_{\text{H}}(\mathbf{x},\mathbf{y})} (1-p)^{2n - d_{\text{H}}(\mathbf{u},\mathbf{v}) - d_{\text{H}}(\mathbf{x},\mathbf{y})}.$$

Proof. We have

$$\begin{aligned} E(P_{\mathrm{ue}}(p)^2) &= \frac{1}{\#\mathcal{C}} \sum_{C\in\mathcal{C}} P_{\mathrm{ue}}(C,p)^2 \\ &= \frac{1}{\#\mathcal{C}} \sum_{C\in\mathcal{C}} \frac{1}{M} \sum_{\substack{(\mathbf{u},\mathbf{v})\in C^2 \\ \mathbf{u}\neq\mathbf{v}}} \Big(\frac{p}{q-1}\Big)^{d_{\mathrm{H}}(\mathbf{u},\mathbf{v})} (1-p)^{n-d_{\mathrm{H}}(\mathbf{u},\mathbf{v})} \\ &\quad \cdot \frac{1}{M} \sum_{\substack{(\mathbf{x},\mathbf{y})\in C^2 \\ \mathbf{x}\neq\mathbf{y}}} \Big(\frac{p}{q-1}\Big)^{d_{\mathrm{H}}(\mathbf{x},\mathbf{y})} (1-p)^{n-d_{\mathrm{H}}(\mathbf{x},\mathbf{y})} \\ &= \frac{1}{M^2} \sum_{\substack{(\mathbf{u},\mathbf{v}),(\mathbf{x},\mathbf{y})\in(F_q^n)^2 \\ \mathbf{u}\neq\mathbf{v},\mathbf{x}\neq\mathbf{y}}} \alpha(\mathbf{u},\mathbf{v},\mathbf{x},\mathbf{y}) \\ &\quad \cdot \Big(\frac{p}{q-1}\Big)^{d_{\mathrm{H}}(\mathbf{u},\mathbf{v})+d_{\mathrm{H}}(\mathbf{x},\mathbf{y})} (1-p)^{2n-d_{\mathrm{H}}(\mathbf{u},\mathbf{v})-d_{\mathrm{H}}(\mathbf{x},\mathbf{y})}. \end{aligned}$$

□

Similarly, we get the following alternative expression for linear codes.

Theorem 2.25. *Let $\mathcal{C}$ be a set of linear codes of length n and dimension k. Then*

$$Var(\mathcal{C},p) = \sum_{\mathbf{x},\mathbf{y}\in GF(q)^n\setminus\{\mathbf{0}\}} \alpha(\mathbf{x},\mathbf{y}) p^{w_H(\mathbf{x})+w_H(\mathbf{y})}(1-p)^{2n-w_H(\mathbf{x})-w_H(\mathbf{y})} - P_{\mathrm{ue}}(\mathcal{C},p)^2.$$

2.4.3 *Average for special classes of codes*

Next, we consider the average for some special sets of codes. First, for some fixed $(n,L;q)$ code K, let

$$\mathcal{C}_{(M)}(K) = \{C \mid C\subseteq K \quad \text{and} \quad \#C = M\}$$

denote the set of $(n,M;q)$ subcodes of K.

Theorem 2.26. *Let K be an $(n,L;q)$ code and $M\le L$. Then*

$$P_{\mathrm{ue}}(\mathcal{C}_{(M)}(K),p) = \frac{M-1}{L-1} P_{\mathrm{ue}}(K,p),$$

Proof. First,

$$\#\mathcal{C}_{(M)}(K) = \binom{L}{M}.$$

We see that if $\mathbf{x} \notin K$ or $\mathbf{y} \notin K$, then $\alpha(\mathbf{x},\mathbf{y}) = 0$. On the other hand, if $\mathbf{x},\mathbf{y} \in K$ and $\mathbf{x} \neq \mathbf{y}$, then

$$\#\{C \in \mathcal{C}_{(M)}(K) \mid \mathbf{x},\mathbf{y} \in C\} = \binom{L-2}{M-2},$$

and so

$$\alpha(\mathbf{x},\mathbf{y}) = \frac{\binom{L-2}{M-2}}{\binom{L}{M}} = \frac{M(M-1)}{L(L-1)},$$

and

$$\begin{aligned} A_i(\mathcal{C}_{(M)}(K)) &= \frac{1}{M} \sum_{\substack{(\mathbf{x},\mathbf{y})\in(Z_q^n)^2 \\ d_H(\mathbf{x},\mathbf{y})=i}} \alpha(\mathbf{x},\mathbf{y}) \\ &= \frac{1}{M} \sum_{\substack{(\mathbf{x},\mathbf{y})\in K^2 \\ d_H(\mathbf{x},\mathbf{y})=i}} \frac{M(M-1)}{L(L-1)} \\ &= \frac{M-1}{L-1} \cdot \frac{1}{L} \sum_{\substack{(\mathbf{x},\mathbf{y})\in K^2 \\ d_H(\mathbf{x},\mathbf{y})=i}} 1 \\ &= \frac{M-1}{L-1} A_i(K). \end{aligned} \tag{2.21}$$

Hence,

$$P_{\text{ue}}(\mathcal{C}_{(M)}(K), p) = \frac{M-1}{L-1} P_{\text{ue}}(K,p).$$

□

Choosing $K = GF(q)^n$ we get the following corollary.

Corollary 2.7. *We have*

$$P_{\text{ue}}(\mathcal{C}_{(M)}(GF(q)^n), p) = \frac{M-1}{q^n-1}\Big\{1-(1-p)^n\Big\}.$$

Remark 2.3. Corollary 2.7 implies in particular that for any given n, M, q, and p there exists an $(n, M; q)$ code C such that

$$P_{\text{ue}}(C,p) \leq \frac{M-1}{q^n-1}\Big\{1-(1-p)^n\Big\}. \tag{2.22}$$

It is an open question if, for given n, M, q, there exists an $(n, M; q)$ code C such that (2.22) is satisfied for all $p \in [0,(q-1)/q]$. A code satisfying (2.22) for all $p \in [0,(q-1)/q]$ is clearly good. However, it is even an open question if there exist good $(n, M; q)$ codes for all q, n and M.

A sufficient condition for the existence of codes satisfying (2.22) is the following.

Theorem 2.27. *Let C be an $(n, M, d; q)$ code with $d > \frac{q-1}{q}n$. Then (2.22) is satisfied for all $p \in [0, (q-1)/q]$.*

Proof. First we note that

$$d \geq \delta = \frac{(q-1)n+1}{q}.$$

By Theorem 2.49 (which is an immediate consequence of Lemma 2.1), we have

$$P_{\mathrm{ue}}(C,p) \leq (M-1)\Big(\frac{p}{q-1}\Big)^{\delta}(1-p)^{n-\delta}. \tag{2.23}$$

Let

$$f(p) = \frac{M-1}{q^n-1}\Big\{1-(1-p)^n\Big\} - (M-1)\Big(\frac{p}{q-1}\Big)^{\delta}(1-p)^{n-\delta}.$$

We see that if $f(p) \geq 0$, then by (2.23), (2.22) is satisfied. We have

$$\begin{aligned} f'(p) &= \frac{(M-1)n}{q^n-1}(1-p)^{n-1} - \frac{M-1}{(q-1)^{\delta}}p^{\delta-1}(1-p)^{n-\delta-1}(\delta-np) \\ &= (M-1)(1-p)^{n-1}\Big\{\frac{n}{q^n-1} - \frac{1}{(q-1)^{\delta}}g(p)\Big\}, \end{aligned}$$

where

$$g(p) = \frac{p^{\delta-1}(\delta-np)}{(1-p)^{\delta}}.$$

We have

$$g'(p) = \frac{\delta p^{\delta-2}}{(1-p)^{\delta+1}}\Big(\delta-1-(n-1)p\Big) = \frac{\delta p^{\delta-2}}{(1-p)^{\delta+1}}(n-1)\Big(\frac{q-1}{q}-p\Big) \geq 0.$$

Hence, $g(p)$ is increasing on $[0, (q-1)/q]$. Since

$$f'(0) = \frac{(M-1)n}{q^n-1} > 0 \text{ and } f'\Big(\frac{q-1}{q}\Big) = \frac{M-1}{q^{n-1}}\Big(\frac{n}{q^n-1} - \frac{1}{q-1}\Big) < 0,$$

this shows that $f'(p)$ is first positive, then negative and so $f(p)$ is first increasing, then decreasing on $[0, (q-1)/q]$. Since $f(0) = f((q-1)/q) = 0$, we can conclude that $f(p) \geq 0$ for all $p \in [0, (q-1)/q]$. This completes the proof. □

Remark 2.4. By Theorem 2.15 we know that $d > \frac{q-1}{q}n$ also implies that C is proper. A simple example which shows that for $d \le \frac{q-1}{q}n$, C may be proper without (2.22) being satisfied, is the proper $(5, 4; 2)$ code

$$C = \{(00000), (00001), (00110), (01111)\}.$$

The distance distribution for this code is $1, 1/2, 1, 1, 1/2, 0$ and it is easy to show that

$$\frac{1}{2}p(1-p)^4 + p^2(1-p)^3 + p^3(1-p)^2 + \frac{1}{2}p^4(1-p) > \frac{3}{31}(1-(1-p)^3)$$

for all $p \in (0, 1/2)$, that is, (2.22) is never satisfied for this code.

We get similar results for the average of linear codes. For some fixed $[n, \kappa; q]$ code K, let

$$\mathcal{C}_{[k]}(K) = \{C \mid C \subseteq K \quad \text{and} \quad \dim(C) = k\}$$

denote the set of $[n, k; q]$ subcodes of K.

Theorem 2.28. *Let K be an $[n, \kappa; q]$ code and $k \le \kappa$. Then*

$$P_{\text{ue}}(\mathcal{C}_{[k]}(K), p) = \frac{q^k - 1}{q^\kappa - 1} P_{\text{ue}}(K, p).$$

Proof. First,

$$\#\mathcal{C}_{[k]}(K) = \begin{bmatrix} \kappa \\ k \end{bmatrix} = \frac{\prod_{i=0}^{k-1}(q^\kappa - q^i)}{\prod_{i=0}^{k-1}(q^k - q^i)},$$

the *Gaussian binomial coefficient.* We see that if $\mathbf{x} \notin K$, then $\alpha(\mathbf{x}) = 0$. On the other hand, if $\mathbf{x} \in K$, then

$$\#\{C \in \mathcal{C}_{[k]}(K) \mid \mathbf{x} \in C\} = \begin{bmatrix} \kappa - 1 \\ k - 1 \end{bmatrix},$$

and so

$$\alpha(\mathbf{x}) = \frac{\begin{bmatrix} \kappa-1 \\ k-1 \end{bmatrix}}{\begin{bmatrix} \kappa \\ k \end{bmatrix}} = \frac{q^k - 1}{q^\kappa - 1},$$

and

$$A_i(\mathcal{C}_{[k]}(K)) = \frac{q^k - 1}{q^\kappa - 1} A_i(K). \tag{2.24}$$

Hence,

$$P_{\text{ue}}(\mathcal{C}_{[k]}(K), p) = \frac{q^k - 1}{q^\kappa - 1} P_{\text{ue}}(K, p).$$

□

Choosing $K = GF(q)^n$ we get the following corollary.

Corollary 2.8. *We have*

$$P_{\rm ue}(\mathcal{C}_{[k]}(GF(q)^n), p) = \frac{q^k - 1}{q^n - 1}\Big\{1 - (1-p)^n\Big\}.$$

Comparing Corollaries 2.7 and 2.8, we see that

$$P_{\rm ue}(\mathcal{C}_{[k]}(GF(q)^n), p) = P_{\rm ue}(\mathcal{C}_{(q^k)}(GF(q)^n), p). \tag{2.25}$$

Even if the averages happen to be the same, there is no reason why, for example, the variances should be the same. This is illustrated by the next example where the variances are different.

Example 2.11. In Example 2.3 we gave a listing of the possible distance distributions of $(5, 4; 2)$ codes. From these we can compute the average and variance. For the average we get

$$P_{\rm ue}(\mathcal{C}_{(4)}(GF(2)^5), p) = P_{\rm ue}(\mathcal{C}_{[2]}(GF(2)^5), p) = \frac{3}{31}\Big\{1 - (1-p)^5\Big\},$$

as we should by (2.25). For the variance we get

$$Var(\mathcal{C}_{(4)}(GF(2)^5), p) = \frac{1}{899}\sum_{i=2}^{10} u_i p^i (1-p)^{10-i},$$

$$Var(\mathcal{C}_{[2]}(GF(2)^5), p) = \frac{1}{31}\sum_{i=2}^{10} v_i p^i (1-p)^{10-i},$$

where the u_i and v_i are given in Table 2.8. It turns out that $Var(\mathcal{C}_{(4)}(GF(2)^5), p) < Var(\mathcal{C}_{[2]}(GF(2)^5), p)$ for all $p \in (0, 1/2)$.

Table 2.8 The coefficients u_i and v_i in Example 2.11.

i	2	3	4	5	6	7	8	9	10
u_i	369	720	1818	1800	1890	864	513	72	45
v_i	19	20	68	50	70	24	23	2	3

2.4.4 *Average for systematic codes*

Next we will consider the average probability of undetected error for some classes of systematic codes (the variance can be found in a similar way). Let $\mathcal{SYS}(n, k)$ be the set of systematic $(n, q^k; q)$ codes, $\mathcal{SYSL}(n, k)$ be the set of systematic linear $[n, k; q]$ codes and $\mathcal{SYSL}(n, k, d)$ the set of systematic linear $[n, k, d; q]$ codes. We consider first general systematic codes. Then we give a more detailed analysis of some classes of systematic linear codes.

Theorem 2.29. *For $1 \le k \le n$, we have*

$$P_{\text{ue}}(\mathcal{SYS}(n,k), p) = q^{k-n}\Big\{1 - (1-p)^k\Big\}.$$

Proof. We can write the vectors in the form $(\mathbf{u}|\mathbf{v})$, where $\mathbf{u} \in F_q^k$ and $\mathbf{v} \in F_q^{n-k}$. Let C be a systematic $(n, q^k; q)$ code. By definition, two distinct code words of C must differ in some of the first k positions. Hence, for the class $\mathcal{SYS}(n,k)$, we get $\alpha(\mathbf{x}, \mathbf{y}) = 0$ if $\mathbf{x}$ and $\mathbf{y}$ are identical in the first k positions. To get a systematic code, the last $n-k$ positions can be chosen arbitrarily. Hence $\#\mathcal{SYS}(n,k) = \left(q^{n-k}\right)^{q^k}$. On the other hand, if $\mathbf{x}$ and $\mathbf{y}$ differ in some of the first positions, then they are contained in $\left(q^{n-k}\right)^{q^k-2}$ codes. Hence

$$\alpha(\mathbf{x}, \mathbf{y}) = \frac{\left(q^{n-k}\right)^{q^k-2}}{\left(q^{n-k}\right)^{q^k}} = q^{2k-2n}. \tag{2.26}$$

From Theorem 2.22, we get

$$\begin{aligned}
P_{\text{ue}}(\mathcal{C}, p) &= \frac{1}{q^k} \sum_{\substack{((\mathbf{u}|\mathbf{v}),(\mathbf{u}'|\mathbf{v}')) \in (Z_q^n)^2 \\ \mathbf{u} \neq \mathbf{u}'}} q^{2k-2n} \Big(\frac{p}{q-1}\Big)^{d_{\text{H}}(\mathbf{u},\mathbf{u}') + d_{\text{H}}(\mathbf{v},\mathbf{v}')} \\
&\qquad \cdot (1-p)^{n - d_{\text{H}}(\mathbf{u},\mathbf{u}') - d_{\text{H}}(\mathbf{v},\mathbf{v}')} \\
&= q^{k-2n} \sum_{\substack{(\mathbf{u},\mathbf{u}') \in (Z_q^k)^2 \\ \mathbf{u} \neq \mathbf{u}'}} \Big(\frac{p}{q-1}\Big)^{d_{\text{H}}(\mathbf{u},\mathbf{u}')} (1-p)^{k - d_{\text{H}}(\mathbf{u},\mathbf{u}')} \\
&\qquad \cdot \sum_{(\mathbf{v},\mathbf{v}') \in (Z_q^{n-k})^2} \Big(\frac{p}{q-1}\Big)^{d_{\text{H}}(\mathbf{v},\mathbf{v}')} (1-p)^{n-k-d_{\text{H}}(\mathbf{v},\mathbf{v}')} \\
&= q^{k-2n} \sum_{\mathbf{u} \in Z_q^k} \Big\{1 - (1-p)^k\Big\} \sum_{\mathbf{v} \in Z_q^{n-k}} 1 \\
&= q^{k-2n} \cdot q^k \Big\{1 - (1-p)^k\Big\} \cdot q^{n-k} \\
&= q^{k-n}\Big\{1 - (1-p)^k\Big\}.
\end{aligned}$$

□

Theorem 2.30. *We have*

$$A_i(\mathcal{SYSL}(n,k,d)) = 0 \quad \text{for } 1 \le i \le d-1,$$
$$A_i(\mathcal{SYSL}(n,k,d)) \le \frac{(q-1)^i}{q^{n-k} - \sum_{l=0}^{d-2}\binom{n-1}{l}(q-1)^l}\Big\{\binom{n}{i} - \binom{n-k}{i}\Big\},$$

with equality for $d = 1$.

Proof. Let $i \ge d$. A vector of the form $(\mathbf{0}|\mathbf{v})$, where $\mathbf{v} \in GF(q)^{n-k} \setminus \{\mathbf{0}\}$, is not contained in any systematic code. Hence the number of vectors $\mathbf{x}$ of weight i which can be contained in a systematic code is

$$\binom{n}{i}(q-1)^i - \binom{n-k}{i}(q-1)^i.$$

Let $\mathbf{x}$ be such a vector. It remains to show that

$$\alpha(\mathbf{x}) \le \frac{1}{q^{n-k} - \sum_{l=0}^{d-2}\binom{n-1}{l}(q-1)^l}. \tag{2.27}$$

The vector $\mathbf{x}$ contains at least one non-zero element in the first k positions. Without loss of generality, we may assume that it is a 1 in the first position, i.e. $\mathbf{x} = (1|\mathbf{z}|\mathbf{v})$, where $\mathbf{z} \in GF(q)^{k-1}$. Let

$$\begin{pmatrix} 1 & \mathbf{z} & \mathbf{v} \\ \mathbf{0}^{\mathrm{t}} & I_{k-1} & P \end{pmatrix}$$

be a generator matrix for a code in $\mathcal{SYSL}(n,k,d)$ which contains $\mathbf{x}$. Then $(I_{k-1}|P)$ generates a code in $\mathcal{SYSL}(n-1,k-1,d)$. Hence

$$\alpha(\mathbf{x}) \le \frac{\#\mathcal{SYSL}(n-1,k-1,d)}{\#\mathcal{SYSL}(n,k,d)}. \tag{2.28}$$

On the other hand, if $G = (I_{k-1}|P)$ is the generator matrix for a code in $\mathcal{SYSL}(n-1,k-1,d)$, then any non-zero linear combination of $d-1$ or less columns in the corresponding check matrix $(P^{\mathrm{t}}|I_{n-k})$ has a non-zero sum. Let $\mathbf{y}$ be a non-zero vector in $GF(q)^{n-k}$ different from all sums of $d-2$ or less of these columns; $\mathbf{y}$ can be chosen in at least

$$q^{n-k} - 1 - \sum_{l=1}^{d-2}\binom{n-1}{l}(q-1)^l$$

distinct ways. Each choice of $\mathbf{y}$ gives a check matrix $(\mathbf{y}^{\rm t}|P^{\rm t}|I_{n-k})$ such that any combination of $d-1$ or less columns has a non-zero sum, that is, a check matrix for a code in $\mathcal{SYSL}(n,k,d)$. Therefore,

$$\frac{\#\mathcal{SYSL}(n,k,d)}{\#\mathcal{SYSL}(n-1,k-1,d)} \geq q^{n-k} - \sum_{l=0}^{d-2}\binom{n-1}{l}(q-1)^l. \tag{2.29}$$

Combining (2.28) and (2.29), we get (2.27). For $d=1$ we get equality in both (2.28) and (2.29). □

Theorem 2.31. *Let $\mathcal{C} = \mathcal{SYSL}(n,k,d)$, the set of all systematic $[n,k,d;q]$ codes. Then*

$$P_{\rm ue}(\mathcal{C},p) \leq \frac{1-(1-p)^k - \sum_{i=1}^{d-1}\left\{\binom{n}{i}-\binom{n-k}{i}\right\}p^i(1-p)^{n-i}}{q^{n-k} - \sum_{l=0}^{d-2}\binom{n-1}{l}(q-1)^l}$$

with equality for $d=1$.

Proof. Let

$$\beta = \frac{1}{q^{n-k} - \sum_{l=0}^{d-2}\binom{n-1}{l}(q-1)^l}.$$

By Theorem 2.30 we get

$$\begin{aligned} P_{\rm ue}(\mathcal{C},p) &\leq \beta\Big\{\sum_{i=d}^{n}\binom{n}{i}p^i(1-p)^{n-i} - \sum_{i=d}^{n-k}\binom{n-k}{i}p^i(1-p)^{n-i}\Big\} \\ &= \beta\Big\{1-(1-p)^k - \sum_{i=1}^{d-1}\Big\{\binom{n}{i}-\binom{n-k}{i}\Big\}p^i(1-p)^{n-i}\Big\}. \end{aligned}$$

□

From Theorems 2.29 and 2.31 we get

$$P_{\rm ue}(\mathcal{SYSL}(n,k),p) = P_{\rm ue}(\mathcal{SYS}(n,k),p),$$

that is, the average over all systematic linear codes is the same as the average over all systematic codes (when q is a prime power). We had a similar result for the set of all $[n,k;q]$ codes and all $(n,q^k;q)$ codes. We also note that on average, systematic codes are better than codes in general. We state a somewhat stronger theorem.

Theorem 2.32. *Let $1 \le k < n$ and $p \in (0, (q-1)/q)$. The ratio*

$$\frac{P_{\mathrm{ue}}(\mathcal{SYS}(n,k),p)}{P_{\mathrm{ue}}(\mathcal{C}_{(q^k)}(F_q^n),p)} = \frac{q^{k-n}\Big\{1-(1-p)^k\Big\}}{\frac{q^k-1}{q^n-1}\Big\{1-(1-p)^n\Big\}}$$

is increasing on $(0, (q-1)/q)$. For $p \to 0+$ the ratio is $\frac{k(1-q^{-n})}{n(1-q^{-k})}$, and for $p = (q-1)/q$ the ratio is 1.

Proof. Let

$$f(p) = P_{\mathrm{ue}}(\mathcal{SYS}(n,k),p) = q^{k-n}\Big\{1-(1-p)^k\Big\},$$

$$g(p) = P_{\mathrm{ue}}(\mathcal{C}_{(q^k)}(F_q^n),p) = \frac{q^k-1}{q^n-1}\Big\{1-(1-p)^n\Big\}$$

$$h(p) = f(p)/g(p).$$

Then

$$\begin{aligned} g(p)^2 h'(p) &= f(p)g'(p) - f'(p)g(p) \\ &= q^{k-n}\Big\{1-(1-p)^k\Big\}\frac{q^k-1}{q^n-1}n(1-p)^{n-1} \\ &\quad -q^{k-n}k(1-p)^{k-1}\frac{q^k-1}{q^n-1}\Big\{1-(1-p)^n\Big\} \\ &= q^{k-n}\frac{q^k-1}{q^n-1}(1-p)^{k-1}F(p), \end{aligned}$$

where

$$\begin{aligned} F(p) &= n\Big\{1-(1-p)^k\Big\}(1-p)^{n-k} - k\Big\{1-(1-p)^n\Big\} \\ &= n(1-p)^{n-k} - k - (n-k)(1-p)^n. \end{aligned}$$

We have $F(0) = 0$ and

$$\begin{aligned} F'(p) &= -n(n-k)(1-p)^{n-k-1} + (n-k)n(1-p)^{n-1} \\ &= n(n-k)(1-p)^{n-k-1}\Big\{1-(1-p)^k\Big\} > 0 \end{aligned}$$

for all $p \in (0, (q-1)/q)$. Hence $h'(p) > 0$ for all $p \in (0, (q-1)/q)$. □

How good is the upper bound in Theorem 2.31? It is exact for $d = 1$. We will next derive the exact value of $P_{\mathrm{ue}}(\mathcal{SYSL}(n,k,2),p)$ and compare this with the upper bound in Theorem 2.31.

Theorem 2.33. *Let $\mathcal{C} = \mathcal{SYSL}(n,k,2)$. For $1 \le i \le n$ we have*

$$A_i(\mathcal{C}) = \frac{(q-1)^i}{q^r}\Big\{\binom{n}{i} - \binom{k}{i}\varsigma^{i-1} - \sum_{l=0}^{i-1}\binom{k}{l}\binom{r}{i-l}\varsigma^l\Big\},$$

where $r = n - k$ *and* $\zeta = -1/(q^r - 1)$, *and*

$$P_{\text{ue}}(\mathcal{C}, p) = \frac{1}{q^r}\Big\{1 - \delta(p)^k - q^r(1-p)^n + q^r(1-p)^r\delta(p)^k\Big\},$$

where

$$\delta(p) = 1 - \frac{q^r p}{q^r - 1}.$$

Proof. We see that $(I_k|P)$ is a generator matrix for a code in $\mathcal{C}$ if and only if all the rows of P are non-zero. In particular, $\#\mathcal{C} = (q^r - 1)^k$.

For given non-zero elements $\alpha_1, \alpha_2, \ldots \alpha_i$, consider representations of $\mathbf{v} \in GF(q)^r$:

$$\mathbf{v} = \alpha_1\mathbf{x}_1 + \alpha_2\mathbf{x}_2 + \cdots + \alpha_i\mathbf{x}_i \tag{2.30}$$

where $\mathbf{x}_j \in GF(q)^r \setminus \{\mathbf{0}\}$ for $j = 1, 2, \ldots, i$. The number of such representations depends on whether $\mathbf{v}$ is the all zero vector or not. Let $\beta(i)$ be the number of representations of $\mathbf{0} \in GF(q)^r$ and $\gamma(i)$ be the number of representations of $\mathbf{v} \in GF(q)^r \setminus \{\mathbf{0}\}$.

We note that (2.30) is equivalent to

$$\mathbf{v} - \alpha_i\mathbf{x}_i = \alpha_1\mathbf{x}_1 + \alpha_2\mathbf{x}_2 + \cdots + \alpha_{i-1}\mathbf{x}_{i-1}. \tag{2.31}$$

If $\mathbf{v} = \mathbf{0}$, then the left-hand side of (2.31) is non-zero. Hence in this case $(\mathbf{x}_1, \mathbf{x}_2, \ldots, \mathbf{x}_{i-1})$ can be chosen in $\gamma(i-1)$ ways for each of the $q^r - 1$ choices of $\mathbf{x}_i$. Hence

$$\beta(i) = (q^r - 1)\gamma(i-1). \tag{2.32}$$

Similarly, if $\mathbf{v} \neq \mathbf{0}$, then the left-hand side of (2.31) is non-zero except when $\alpha_i\mathbf{x}_i = \mathbf{v}$. Hence we get

$$\beta(i) = (q^r - 2)\gamma(i-1) + \beta(i-1). \tag{2.33}$$

The pair of equations (2.32) and (2.33) constitutes a simple linear recursion that can be solved by standard methods. The start of the recursion is the obvious values $\beta(1) = 0$ and $\gamma(1) = 1$. The result, that can easily be verified, is

$$\beta(i) = \frac{q^r - 1}{q^r}\Big\{(q^r - 1)^{i-1} - (-1)^{i-1}\Big\},$$

$$\gamma(i) = \frac{1}{q^r}\Big\{(q^r - 1)^i - (-1)^i\Big\}.$$

The number of codes in $\mathcal{C}$ containing a vector $(\mathbf{u}|\mathbf{0})$ where $w_H(\mathbf{u}) = i$ is $\beta(i)(q^r - 1)^{k-i}$ and the number of such vectors is $\binom{k}{i}(q-1)^i$. For $0 <$

$l < i$, the number of codes containing a vector $(\mathbf{u}|\mathbf{v})$ where $w_H(\mathbf{u}) = l$ and $w_H(\mathbf{v}) = i - l$ is $\gamma(l)(q^r - 1)^{k-l}$, and the number of such vectors is $\binom{k}{l}\binom{r}{i-l}(q-1)^i$. Hence

$$\begin{aligned} A_i(\mathcal{C}) &= \binom{k}{i}\frac{(q-1)^i}{(q^r-1)^i}\cdot\frac{q^r-1}{q^r}\Big\{(q^r-1)^{i-1}-(-1)^{i-1}\Big\} \\ &\quad+\sum_{l=1}^{i-1}\binom{k}{l}\binom{r}{i-l}\frac{(q-1)^i}{(q^r-1)^l}\cdot\frac{1}{q^r}\Big\{(q^r-1)^l-(-1)^l\Big\} \\ &= \frac{(q-1)^i}{q^r}\Big\{\sum_{l=1}^{i}\binom{k}{l}\binom{r}{i-l}-\binom{k}{i}\varsigma^{i-1}-\sum_{l=1}^{i-1}\binom{k}{l}\binom{r}{i-l}\varsigma^l\Big\}. \end{aligned}$$

Since

$$\sum_{l=1}^{i}\binom{k}{l}\binom{r}{i-l}=\binom{n}{i}-\binom{r}{i}$$

we get the expression for A_i given in the theorem. Putting this into the expression $P_{\text{ue}}(\mathcal{C},p) = \sum_{i=1}^{n} A_i(\mathcal{C})\left(\frac{p}{q-1}\right)^i(1-p)^{n-i}$ and rearranging, we get the expression for $P_{\text{ue}}(\mathcal{C},p)$ given in the theorem. □

The upper bound in Theorem 2.31 is exact for $p = 0$ and too large by the quantity $\frac{q^k-k-1}{(q^r-1)q^n}$ for $p = \frac{q-1}{q}$. The difference between the upper bound and the exact value appears to increase monotonously when p increases from 0 to $\frac{q-1}{q}$. The exact value of $P_{\text{ue}}(\mathcal{SYSL}(n,k,d),p)$ for general d is not known, and it is probably quite complicated both to derive and express.

Our final example is systematic binary even weight codes.

Theorem 2.34. *Let $\mathcal{C}$ be the set of all systematic even weight $[n,k;2]$ codes. Then*

$$A_i(\mathcal{C}) = 0 \quad \textit{for odd } i,$$
$$A_i(\mathcal{C}) = \frac{1}{2^{r-1}}\Big\{\binom{n}{i}-\binom{n-k}{i}\Big\} \quad \textit{for even } i > 0,$$

and

$$P_{\text{ue}}(\mathcal{C},p) = \frac{1}{2^{n-k}}\Big\{1+(1-2p)^n-(1-p)^k-(1-2p)^{n-k}(1-p)^k\Big\}.$$

Proof. The proof is similar, and we only give a sketch. All the rows of P have to have odd weight. For a vector $\mathbf{x} = (\mathbf{u}|\mathbf{v})$, where $\mathbf{u} \neq \mathbf{0}$ and $w_H(\mathbf{x})$ is even, we have

$$\alpha(\mathbf{x}) = 2^{-(r-1)},$$

and for even i there are $\binom{n}{i} - \binom{n-k}{i}$ such vectors of weight i. This proves the expression for $A_i(\mathcal{C})$. Further,

$$\begin{aligned} P_{\mathrm{ue}}(\mathcal{C},p) &= 2^{-(r-1)} \sum_{i=1}^{\lfloor n/2 \rfloor} \left\{ \binom{n}{2i} - \binom{n-k}{2i} \right\} p^{2i}(1-p)^{n-2i} \\ &= 2^{-r}\Big\{1 + (1-2p)^n - (1-p)^k - (1-2p)^r(1-p)^k\Big\}. \end{aligned}$$

□

2.5 The worst-case error probability

In many applications we do not know the value of p, at least not exactly. In other applications we may want to use one code for several different values of p. Therefore, we will be interested in finding the *worst-case error probability*

$$P_{\mathrm{wc}}(C,a,b) = \max_{a \le p \le b} P_{\mathrm{ue}}(C,p)$$

for some interval $[a,b] \subseteq [0,1]$. First we give an upper bound.

Theorem 2.35. *Let C be an $(n, M; q)$ code. Then*

$$P_{\mathrm{wc}}(C,a,b) \le \sum_{i=1}^{n} A_i \mu_{n,i}(a,b) \tag{2.34}$$

where $A_0, A_1, \ldots, A_n$ is the distance distribution of C and

$$\mu_{n,i}(a,b) = \begin{cases} \left(\frac{a}{q-1}\right)^i (1-a)^{n-i} & \text{if } \frac{i}{n} < a, \\ \left(\frac{i}{n(q-1)}\right)^i \left(1 - \frac{i}{n}\right)^{n-i} & \text{if } a \le \frac{i}{n} \le b, \\ \left(\frac{b}{q-1}\right)^i (1-b)^{n-i} & \text{if } \frac{i}{n} > b. \end{cases}$$

If C is a constant distance code, we have equality in (2.34).

Proof. First we note that the function $p^i(1-p)^{n-i}$ is increasing on the interval $\left[0, \frac{i}{n}\right]$ and decreasing on the interval $\left[\frac{i}{n}, 1\right]$. Hence

$$\max_{a \le p \le b} \left\{ \left(\frac{p}{q-1}\right)^i (1-p)^{n-i} \right\} = \mu_{n,i}(a,b),$$

and so

$$\begin{aligned}P_{\rm wc}(C,a,b) &= \max_{a\le p\le b}\sum_{i=1}^{n} A_i\Big(\frac{p}{q-1}\Big)^i(1-p)^{n-i}\\ &\le \sum_{i=1}^{n} A_i \max_{a\le p\le b}\Big(\frac{p}{q-1}\Big)^i(1-p)^{n-i}\\ &= \sum_{i=1}^{n} A_i\mu_{n,i}(a,b).\end{aligned}$$

If C is a constant distance code we get equality. □

The bound in Theorem 2.35 may be improved by subdividing the interval $[a,b]$: let $a = a_0 < a_1 < \cdots < a_m = b$. Then clearly

$$P_{\rm wc}(C,a,b) = \max_{1\le j\le m} P_{\rm wc}(C,a_{j-1},a_j).$$

It is easy to see that if

$$F = \sum_{i=1}^{n} A_i \max_{a\le p\le b}\Big|\frac{d}{dp}\Big(\Big(\frac{p}{q-1}\Big)^i(1-p)^{n-i}\Big)\Big|,$$

then

$$\begin{aligned}P_{\rm wc}(C,a_{j-1},a_j) &\le \sum_{i=1}^{n} A_i\mu_{n,i}(a_{j-1},a_j)\\ &\le P_{\rm ue}(C,a_{j-1}) + (a_j - a_{j-1})F\\ &\le P_{\rm wc}(C,a_{j-1},a_j) + (a_j - a_{j-1})F.\end{aligned}$$

Hence we can get as sharp a bound as we want by subdividing the interval sufficiently.

Remark 2.5. It is possible to find an upper bound similar to the bound in Theorem 2.35, but in terms of the dual distribution $A_0^\perp, A_1^\perp, \ldots, A_n^\perp$:

$$P_{\rm wc}(C,a,b) \le \frac{1}{q^{n-k}}\sum_{i=0}^{n} A_i^\perp \nu_{n,i}(a,b)$$

where $\nu_{n,i}(a,b)$ is the maximum of the function $f_{n,i}(p) = \Big(1 - \frac{q}{q-1}p\Big)^i - (1-p)^n$ on the interval $[a,b]$. As in the proof above, the maximum is obtained in a, in b or in an inner point in (a,b). In most cases we cannot find explicit expressions for the maxima of $f_{n,i}(p)$, but we can find as good approximations as we like. If we only want to find the worst-case of one code on one interval, it is probably a better idea to do this directly. However, once we have computed the maxima of $f_{n,i}$ for $i = 0, 1, \ldots, n$, then it is a simple task to compute $\nu_{n,i}(a, b)$ for any a and b, and since this is independent of the code, we can compute the worst-case probability of any code for which we know the dual distribution (of course, an alternative is to compute the distance distribution of the code itself using MacWilliams's identity and then using Theorem 2.35. However, it may be simpler in some cases to use the attack sketched above).

Next we consider a bound on the average value of $P_{\text{wc}}(C, a, b)$ over some class $\mathcal{C}$. Let

$$P_{\text{wc}}(\mathcal{C}, a, b) = \frac{1}{\#\mathcal{C}} \sum_{c \in \mathcal{C}} P_{\text{wc}}(C, a, b).$$

From Theorem 2.35 we immediately get the following bound.

Theorem 2.36. *Let $\mathcal{C}$ be a class of $(n, M; q)$ codes. Then*

$$P_{\text{wc}}(\mathcal{C}, a, b) \le \sum_{i=1}^{n} A_i(\mathcal{C}) \mu_{n,i}(a, b).$$

We now consider the worst-case over $[0, 1]$. Since

$$\mu_{n,i}(0, 1) = \Big(\frac{i}{n(q-1)}\Big)^i \Big(1 - \frac{i}{n}\Big)^{n-i},$$

we get the following corollary to Theorem 2.36.

Corollary 2.9. *For a set of $(n, M; q)$ codes $\mathcal{C}$ we have*

$$P_{\text{wc}}(\mathcal{C}, 0, 1) \le \sum_{i=1}^{n} A_i(\mathcal{C}) \Big(\frac{i}{n(q-1)}\Big)^i \Big(1 - \frac{i}{n}\Big)^{n-i}.$$

Some special cases of Corollary 2.9 are given in the following corollaries, obtained using (2.21) and Theorem 2.30.

Corollary 2.10. *We have*

$$P_{\text{wc}}(\mathcal{C}_{(M)}(GF(q)^n), 0, 1) \le \frac{M-1}{q^n - 1} S_n,$$

where

$$S_n = \sum_{i=1}^{n} \binom{n}{i} \Big(\frac{i}{n}\Big)^i \Big(1 - \frac{i}{n}\Big)^{n-i}. \tag{2.35}$$

Corollary 2.11. *We have*

$$P_{\rm wc}(\mathcal{SYSL}(n,k),0,1) \le \frac{1}{q^{n-k}}(S_n - S_{n,k}),$$

where

$$S_{n,k} = \sum_{i=1}^{n} \binom{n-k}{i} \Big(\frac{i}{n}\Big)^i \Big(1-\frac{i}{n}\Big)^{n-i}. \tag{2.36}$$

Note that S_n and $S_{n,k}$ do not depend on q. Some relations for the sums S_n and $S_{n,k}$ are the following:

$$\sqrt{\frac{\pi n}{2}} - \frac{q-1}{q} < S_n < \sqrt{\frac{\pi n}{2}} - \frac{1}{3} + \frac{0.108918}{\sqrt{n}}, \tag{2.37}$$

$$S_n = \sqrt{\frac{\pi n}{2}} - \frac{1}{3} + \frac{\sqrt{2\pi}}{24\sqrt{n}} + O\Big(\frac{1}{n}\Big) \quad \text{when} \quad n \to \infty, \tag{2.38}$$

$$S_{n,k} = \frac{1}{q^{2k}} \binom{2k}{k} \sqrt{\frac{\pi n}{2}} - \frac{2}{3} + \Big(\frac{1}{\sqrt{n}}\Big) \quad \text{when} \quad n \to \infty, \tag{2.39}$$

for $k \ge 1$,

$$S_{n,1} = \frac{q-1}{q} S_n. \tag{2.40}$$

For example,

$$P_{\rm wc}(\mathcal{C}_{[1]}(GF(q)^n),0,1) = \frac{1}{q^n-1} S_n,$$

and

$$P_{\rm wc}(\mathcal{SYSL}(n,1),0,1) = \frac{1}{q^n} S_n.$$

For $q = 2$ and the worst-case over $[0, 1/2]$ we get

$$\mu_{n,i}\Big(0,\frac{1}{2}\Big) = \Big(\frac{i}{n}\Big)^i \Big(1-\frac{i}{n}\Big)^{n-i} \quad \text{for} \quad i \le \frac{n}{2},$$
$$\mu_{n,i}\Big(0,\frac{1}{2}\Big) = \frac{1}{2^n} \quad \text{for} \quad i \ge \frac{n}{2}.$$

Since

$$\sum_{i=1}^{\lceil n/2 \rceil} \Big(\frac{i}{n}\Big)^i \Big(1-\frac{i}{n}\Big)^{n-i} + \sum_{\lceil n/2 \rceil + 1}^{n} \frac{1}{2^n} = \frac{1}{2} S_n$$

we get

$$P_{\rm wc}\Big(\mathcal{C}_{[k]}(GF(q)^n),0,\frac{1}{2}\Big) \le \frac{2^k-1}{2(2^n-1)} S_n. \tag{2.41}$$

In particular, for given n and k, there exists an $[n, k; 2]$ code C such that

$$P_{\rm wc}\Big(C, 0, \frac{1}{2}\Big) < \frac{2^k - 1}{2(2^n - 1)}\Big(\sqrt{\frac{\pi n}{2}} - \frac{1}{3} + \frac{0.108918}{\sqrt{n}}\Big). \tag{2.42}$$

The following improvements have been shown.

Theorem 2.37. *There is a constant γ such that for given n and k, there exists an $[n, k; 2]$ code C such that*

$$P_{\rm wc}\Big(C, 0, \frac{1}{2}\Big) \le 2^{k-n}(\gamma\sqrt{\ln n} + 1).$$

For sufficiently large n, the result is true for $\gamma = 2/\sqrt{n}$.

Theorem 2.38. *There are constants ν and γ_1 such that if $k > \nu(\ln n)^2$, there exists an $[n, k; 2]$ code C such that*

$$P_{\rm wc}\Big(C, 0, \frac{1}{2}\Big) \le \gamma_1 2^{k-n}.$$

Corollary 2.10 shows that an average code is not too bad. However, some codes may be very bad as shown by the next theorem.

Theorem 2.39. *For each $\gamma > 0$ and each $\epsilon \in \Big(0, \frac{q-1}{2q}\Big)$, for all*

$$n \ge \max\Big(\frac{1}{\varepsilon}, 1 + \frac{\gamma}{q\varepsilon\{1 - q\varepsilon/(q-1)\}}\Big)$$

there exists a code C of length n such that

$$P_{\rm ue}(C, p) > \gamma \cdot P_{\rm ue}\Big(C, \frac{q-1}{q}\Big)$$

for all $p \in \Big[\epsilon, \frac{q-1}{q} - \epsilon\Big]$.

Proof. Let $C_n = \{(a|\mathbf{0}) \in F_q^n \mid a \in F_q\}$. Then

$$P_{\rm ue}(C_n, p) = f_n(p) = (q-1)\frac{p}{q-1}(1-p)^{n-1} = p(1-p)^{n-1},$$

and

$$\frac{d}{dp} f_n(p) = (1-p)^{n-2}(1 - np).$$

Hence, if $n > \frac{1}{\epsilon}$, then $1 - np < 0$ for $p \ge \varepsilon$, and so $f_n(p)$ is decreasing on $[\varepsilon, 1]$. Further, if $n \ge 1 + \frac{\gamma}{q\varepsilon\{1 - q\varepsilon/(q-1)\}}$, then

$$\frac{f_n\Big(\frac{q-1}{q} - \epsilon\Big)}{f_n\Big(\frac{q-1}{q}\Big)} = \Big(1 - \frac{q}{q-1}\epsilon\Big)(1 + q\epsilon)^{n-1} \ge \Big(1 - \frac{q}{q-1}\epsilon\Big)q\varepsilon(n-1) \ge \gamma.$$

□

Example 2.12. For $q = 2$, $N = 4$ and $\epsilon = 0.1$, we can choose any $n \ge 10$.

2.6 General bounds

Define

$$P_{\text{ue}}(n, M, p; q) = \min\{P_{\text{ue}}(C, p) \mid C \text{ is an } (n, M; q) \text{ code}\},$$
$$P_{\text{ue}}[n, k, p; q] = \min\{P_{\text{ue}}(C, p) \mid C \text{ is an } [n, k; q] \text{ code}\}.$$

In many cases one is not able to find the weight distribution. Therefore, we give a number of lower and upper bounds on $P_{\text{ue}}[n, k, p; q]$ and $P_{\text{ue}}(n, M, p; q)$ which may be used to estimate P_{ue}.

2.6.1 *Lower bounds*

We have $A_d \geq 2/M$ for an $(n, M, d; q)$ code and $A_d \geq q - 1$ for an $[n, k, d; q]$ code. Hence we get the following two trivial lower bounds.

Theorem 2.40. *For any $(n, M, d; q)$ code C and any $p \in \left[0, \frac{q-1}{q}\right]$ we have*

$$P_{\text{ue}}(C, p) \geq \frac{2}{M}\left(\frac{p}{q-1}\right)^d (1-p)^{n-d}.$$

Theorem 2.41. *For any $[n, k, d; q]$ code C and any $p \in \left[0, \frac{q-1}{q}\right]$ we have*

$$P_{\text{ue}}(C, p) \geq (q-1)\left(\frac{p}{q-1}\right)^d (1-p)^{n-d}.$$

There are a couple of bounds that follow from Lemma 2.1.

Theorem 2.42. *For any n, k, and any $p \in \left[0, \frac{q-1}{q}\right]$ we have*

$$P_{\text{ue}}[n, k, p; q] \geq \left(q^k - 1\right)\left(\frac{p}{q-1}\right)^{\frac{nq^{k-1}}{q^k-1}} (1-p)^{\frac{n\left(q^{k-1}-1\right)}{q^k-1}}.$$

Proof. Let $C = \{\mathbf{x}_0 = \mathbf{0}, \mathbf{x}_1, \cdots, \mathbf{x}_{q^k-1}\}$ be an $[n, k; q]$ code and let its support weight be m. Let $t_i = w_H(\mathbf{x}_i)$, the Hamming weight of the $\mathbf{x}_i$. Then

$$P_{\text{ue}}(C, p) = \sum_{i=1}^{q^k-1} \left(\frac{p}{q-1}\right)^{t_i} (1-p)^{n-t_i}.$$

If j is in the support of C, then $1/q$ of the code words of C have a zero in position j, the remaining have a non-zero element. Hence

$$\sum_{i=1}^{q^k-1} t_i = mq^{k-1}.$$

Let $\lambda_i = \left(\frac{p}{q-1}\right)^{t_i}(1-p)^{n-t_i}$. Then $P_{\text{ue}}(C,p) = \sum_{i=1}^{q^k-1}\lambda_i$. Moreover

$$\prod_{i=1}^{q^k-1}\lambda_i = \left(\frac{p}{q-1}\right)^{\sum_{i=1}^{q^k-1}t_i}(1-p)^{(q^k-1)n-\sum_{i=1}^{q^k-1}t_i}$$
$$= \left(\frac{p}{q-1}\right)^{mq^{k-1}}(1-p)^{(q^k-1)n-mq^{k-1}}.$$

If $y_1, y_2, \cdots, y_N$ are real numbers such that $\prod_{i=1}^{N} y_i = c > 0$, then it is easily seen that $\sum_{i=1}^{N} y_i \geq Nc^{\frac{1}{N}}$, and the minimum is obtained when all the y_i are equal. In particular, we get

$$P_{\text{ue}}(C,p) \geq (q^k-1)\left(\frac{p}{q-1}\right)^{m\frac{q^{k-1}}{q^k-1}}(1-p)^{n-m\frac{q^{k-1}}{q^k-1}}$$
$$= (q^k-1)(1-p)^n\left(\frac{p}{(q-1)(1-p)}\right)^{m\frac{q^{k-1}}{q^k-1}}$$
$$\geq (q^k-1)(1-p)^n\left(\frac{p}{(q-1)(1-p)}\right)^{n\frac{q^{k-1}}{q^k-1}}$$

for all $p \in \left[0, \frac{q-1}{q}\right]$. □

Theorem 2.43. *For any n, M, and any $p \in \left[0, \frac{q-1}{q}\right]$ we have*

$$P_{\text{ue}}(n, M, p; q) \geq \frac{M}{q^n} - (1-p)^n.$$

Proof. Let C be an $(n, M; q)$ code. Then

$$P_{\text{ue}}(C,p) = \frac{M}{q^n}A_C^{\perp}\left(1 - \frac{q}{q-1}p\right) - (1-p)^n \geq \frac{M}{q^n} - (1-p)^n.$$
□

Remark 2.6. The bound in Theorem 2.43 is negative and hence not interesting for $p < 1 - \frac{q-1}{q}M^{1/n}$. However, for $p \to \frac{q-1}{q}$ it is asymptotically tight.

From Theorem 2.5 and Corollary 1.2 we get the following bound.

Theorem 2.44. *For any n, M, and any $p \in \left[0, \frac{q-1}{q}\right]$ we have*

$$P_{\text{ue}}(n, M, p; q) \geq \sum_{i=n-\lfloor \log_q(M)\rfloor}^{n} \binom{n}{i}\left(\frac{M}{q^{n-i}} - 1\right)\left(\frac{p}{q-1}\right)^i\left(1 - \frac{q}{q-1}p\right)^{n-i}.$$

Using Theorem 1.11 we get a slightly stronger bound.

Theorem 2.45. *For any n, M, and any $p \in \left[0, \frac{q-1}{q}\right]$ we have*

$$P_{\mathrm{ue}}(n, M, p; q) \geq \sum_{i=n-\lfloor \log_q(M) \rfloor}^{n} \alpha_i \Big(\frac{p}{q-1}\Big)^i \Big(1 - \frac{q}{q-1}p\Big)^{n-i},$$

where

$$\alpha_i = \binom{n}{i} \Big(\Big\lceil \frac{M}{q^{n-i}} \Big\rceil - 1\Big) \Big(2 - \frac{q^{n-i}}{M} \Big\lceil \frac{M}{q^{n-i}} \Big\rceil\Big).$$

For linear codes there is a bound of the same form as Theorem 2.43 which follows directly from Corollary 1.10.

Theorem 2.46. *For any n, k, and any $p \in \left[0, \frac{q-1}{q}\right]$ we have*

$$P_{\mathrm{ue}}[n, k, p; q] \geq \frac{1}{q^{n-k}} \prod_{j=k+1}^{n} \Big(1 + (q-1)\Big(1 - \frac{q}{q-1}p\Big)^j\Big) - (1-p)^n.$$

Lemma 2.7. *Let C be an $(n, M; q)$ code, $0 \leq u \leq 1$, and $p \in \left[0, \frac{q-1}{q}\right]$. Then*

$$P_{\mathrm{ue}}(C, p) \geq (M-1)^{1-\frac{1}{u}} \Big\{(q-1)\Big(\frac{p}{q-1}\Big)^u + (1-p)^u\Big\}^{\frac{n}{u}}$$
$$\cdot P_{\mathrm{ue}}\Big(C, \frac{(q-1)p^u}{(q-1)p^u + (q-1)^u(1-p)^u}\Big)^{\frac{1}{u}}.$$

Proof. Let

$$p_u = \frac{(q-1)p^u}{(q-1)p^u + (q-1)^u(1-p)^u}. \tag{2.43}$$

Then

$$\frac{p_u}{(q-1)(1-p_u)} = \Big(\frac{p}{(q-1)(1-p)}\Big)^u.$$

Let $\sum_{i=0}^{n} A_i z^i$ be the distance distribution function of C. Using the Hölder inequality (see e.g. Rudin (1970, page 62)) we get

$$P_{\text{ue}}(C,p_u) = (1-p_u)^n \sum_{i=1}^{n} A_i\Big(\frac{p}{(q-1)(1-p)}\Big)^{ui}$$
$$= \frac{(1-p)^{un}}{\Big\{(q-1)\Big(\frac{p}{q-1}\Big)^u + (1-p)^u\Big\}^n} \sum_{i=1}^{n} \Big\{A_i\Big(\frac{p}{(q-1)(1-p)}\Big)^i\Big\}^u A_i^{1-u}$$
$$\le \frac{(1-p)^{un}}{\Big\{(q-1)\Big(\frac{p}{q-1}\Big)^u + (1-p)^u\Big\}^n} \Big\{\sum_{i=1}^{n} A_i\Big(\frac{p}{(q-1)(1-p)}\Big)^i\Big\}^u \Big\{\sum_{i=1}^{n} A_i\Big\}^{1-u}$$
$$= \frac{1}{\Big\{(q-1)\Big(\frac{p}{q-1}\Big)^u + (1-p)^u\Big\}^n} P_{\text{ue}}(C,p)^u (M-1)^{1-u}.$$

Solving for $P_{\text{ue}}(C,p)$, the theorem follows. □

Theorem 2.47. *Let* $M = \lceil q^{Rn} \rceil$ *and*

$$\alpha(n,R) = -\frac{R}{\log_q(q^{1-R+\frac{1}{n}\log_q(2)} - 1) - \log_q(q-1)}.$$

Then

$$P_{\text{ue}}(n,M,p;q) \ge (M-1)\Big(\frac{p}{q-1}\Big)^{n\alpha(n,R)} (1-p)^{n(1-\alpha(n,R))}.$$

Proof. From the definition of $\alpha = \alpha(n,R)$ we get

$$q^{-R/\alpha} = \frac{q^{1-R}2^{1/n} - 1}{q-1}$$

which in turn implies

$$q^{Rn-n} = 2\frac{1}{(q-1)q^{-R/\alpha} + 1}. \tag{2.44}$$

Let

$$u = -\frac{R}{\alpha \log_q\big(\frac{p}{(q-1)(1-p)}\big)}$$

and define p_u by (2.43). Then

$$q^{-R/\alpha} = \Big(\frac{p}{(q-1)(1-p)}\Big)^u = \frac{p_u}{(q-1)(1-p_u)}$$

and so

$$1 - p_u = \frac{1}{(q-1)q^{-R\alpha} + 1}.$$

By (2.44), this implies that

$$q^{Rn-n} = 2(1-p_u)^n. \tag{2.45}$$

We have $M = \lceil q^{Rn} \rceil \geq q^{Rn} > M-1$ and so

$$\frac{1}{M-1} > q^{-Rn}.$$

Combining all this with Lemma 2.7 and Theorem 2.43 we get

$$\begin{aligned}
P_{\text{ue}}(C,p) &\geq (M-1)^{1-\frac{1}{u}}\Big\{(q-1)\Big(\frac{p}{q-1}\Big)^u + (1-p)^u\Big\}^{\frac{n}{u}} \\
&\quad \cdot \Big\{\frac{M}{q^n} - \Big(\frac{(1-p)^u}{\Big((q-1)\frac{p}{q-1}\Big)^u + (1-p)^u}\Big)^n\Big\}^{\frac{1}{u}} \\
&\geq (M-1)(M-1)^{-\frac{1}{u}}\Big\{(q-1)\Big(\frac{p}{q-1}\Big)^u + (1-p)^u\Big\}^{\frac{n}{u}} \\
&\quad \cdot \Big\{q^{Rn-n} - (1-p_u)^n\Big\}^{\frac{1}{u}} \\
&\geq (M-1)q^{-Rn/u}\Big\{(q-1)\Big(\frac{p}{q-1}\Big)^u + (1-p)^u\Big\}^{\frac{n}{u}}(1-p_u)^{n/u} \\
&= (M-1)\Big(\frac{p}{(q-1)(1-p)}\Big)^{n\alpha}(1-p)^n.
\end{aligned}$$

□

Theorem 2.48. *Let n, $2 \leq K \leq M$ and $p \in \left[0, \frac{q-1}{q}\right]$ be given. Then we have*

$$P_{\text{ue}}(n,M,p;q) \geq \frac{M-K+1}{K-1}\Big(\frac{p}{q-1}\Big)^{d(n,K)}(1-p)^{n-d(n,K)}.$$

Proof. Let C be an (n,M) code and let

$$E = \{(\mathbf{x},\mathbf{y}) \in C^2 \mid 0 < d_H(\mathbf{x},\mathbf{y}) \leq d(n,K)\}.$$

We consider C the vertex set and E the edge set of a graph. Let $F \subseteq C$ be an independent set (no two vertices in F are connected by an edge). Then

$$d_H(\mathbf{x},\mathbf{y}) > d(n,K)$$

for all $\mathbf{x},\mathbf{y} \in F$. By the definition of $d(n,K)$, this implies that $\#F \leq K-1$. A result by Turan (see: Ore (1962, Theorem 13.4.1)) implies that

$$\#E \geq \frac{(M-K+1)M}{K-1}.$$

Hence

$$\begin{aligned}
P_{\text{ue}}(C,p) &\geq \frac{1}{M}\sum_{(\mathbf{x},\mathbf{y})\in E}\Big(\frac{p}{q-1}\Big)^{d_H(\mathbf{x},\mathbf{y})}(1-p)^{n-d_H(\mathbf{x},\mathbf{y})} \\
&\geq \frac{M-K+1}{K-1}\Big(\frac{p}{q-1}\Big)^{d(n,K)}(1-p)^{n-d(n,K)}.
\end{aligned}$$

□

2.6.2 Upper bounds

We now turn to upper bounds. From Lemma 2.1 we immediately get the next theorem.

Theorem 2.49. *For any $(n, M, d; q)$ code C and any $p \in \left[0, \frac{q-1}{q}\right]$ we have*

$$P_{\text{ue}}(C,p) \le \left(M-1\right)\left(\frac{p}{q-1}\right)^d (1-p)^{n-d}.$$

The bound can be sharpened using upper bounds on A_i. The proof is similar, using Lemma 2.2.

Theorem 2.50. *Let C be an $(n, M, d; q)$ code and $p \in \left[0, \frac{q-1}{q}\right]$. Suppose that $A_i(C) \le \alpha_i$ for $d \le i \le n$. Let l be minimal such that*

$$\sum_{i=d}^{l} \alpha_i \ge M - 1.$$

Then

$$P_{\text{ue}}(C,p) \le \sum_{i=d}^{l-1} \alpha_i \left(\frac{p}{q-1}\right)^i (1-p)^{n-i} + \left(M-1-\sum_{i=d}^{l-1}\alpha_i\right)\left(\frac{p}{q-1}\right)^l (1-p)^{n-l}.$$

Example 2.13. Consider an $[8, 4, 3; 2]$ code. By Theorem 2.49 we get

$$P_{\text{ue}}(C,p) \le 15p^3(1-p)^5.$$

Using the bounds on A_i given in Theorem 1.22 we get

$$A_3 \le \left\lfloor \frac{28}{3} \right\rfloor = 9, \quad A_4 \le 14$$

and so

$$P_{\text{ue}}(C,p) \le 9p^3(1-p)^5 + 6p^4(1-p)^4.$$

Theorem 2.51. *Let C be an $[n, k, d; q]$ code and $p \in \left[0, \frac{q-1}{q}\right]$. Then*

$$P_{\text{ue}}(C,p) \le \left(\frac{p}{q-1}\right)^d (1-p)^{n-d} \sum_{i=1}^{d} \binom{k}{i} (q-1)^i + \sum_{i=d+1}^{k} \binom{k}{i} p^i (1-p)^{n-i}.$$

In particular

$$P_{\text{ue}}(C,p) \le (1-p)^{n-k} - (1-p)^n$$

for any $[n, k; q]$ code C.

Proof. Let G be a generator matrix for C. Since equivalent codes have the same probability of undetected error, we may assume without loss of generality that $G = \left(I_k|Q\right)$ where Q is some $k \times (n-k)$ matrix. We have

$$w_H(\mathbf{x}G) = w_H(\mathbf{x}) + w_H(\mathbf{x}Q) \geq w_H(\mathbf{x}).$$

Also $w_H(\mathbf{x}G) \geq d$. Hence

$$\begin{aligned} P_{\text{ue}}(C,p) &= \sum_{\mathbf{x}\in GF(q)^k\setminus\{\mathbf{0}\}} \left(\frac{p}{q-1}\right)^{w_H(\mathbf{x}G)} (1-p)^{n-w_H(\mathbf{x}G)} \\ &\leq \sum_{\substack{\mathbf{x}\in GF(q)^k \\ 1\leq w_H(\mathbf{x})\leq d}} \left(\frac{p}{q-1}\right)^{d} (1-p)^{n-d} \\ &\quad + \sum_{\substack{\mathbf{x}\in GF(q)^k \\ d+1\leq w_H(\mathbf{x})\leq k}} \left(\frac{p}{q-1}\right)^{w_H(\mathbf{x})} (1-p)^{n-w_H(\mathbf{x})}. \end{aligned}$$

In particular

$$\begin{aligned} P_{\text{ue}}(C,p) &\leq (1-p)^{n-k} \sum_{\mathbf{x}\in GF(q)^k\setminus\{\mathbf{0}\}} \left(\frac{p}{q-1}\right)^{w_H(\mathbf{x})} (1-p)^{k-w_H(\mathbf{x})} \\ &= (1-p)^{n-k}\left(1-(1-p)^k\right). \end{aligned}$$

□

Remark 2.7. If C is the code generated by $\left(I_k|0_{k\times(n-k)}\right)$, where $0_{k\times(n-k)}$ is the $k \times (n-k)$ matrix with all entries zero, then

$$P_{\text{ue}}(C,p) = (1-p)^{n-k} - (1-p)^n.$$

Hence, Theorem 2.51 is best possible for $[n,k,1;q]$ codes.

Example 2.14. If we use Theorem 2.51 we get

$$P_{\text{ue}}(C,p) \leq 14p^3(1-p)^5 + p^4(1-p)^4$$

for an $[8,4,3;2]$ code, and this is weaker than the bound obtained from Theorems 1.22 and 2.50. In other cases, Theorem 2.51 may give a better bound. For example, for a $[14,5,3;2]$ code C, Theorem 2.51 gives

$$P_{\text{ue}}(C,p) \leq 25p^3(1-p)^{11} + 5p^4(1-p)^{10} + p^5(1-p)^9,$$

whereas Theorems 1.22 and 2.50 give only

$$P_{\text{ue}}(C,p) \leq 30p^3(1-p)^{11} + p^4(1-p)^{10}.$$

In Example 2.3, we gave the distance distribution of all $(5,4;2)$ codes. As shown in that example, $P_{\text{ue}}(C,p) \le (1-p)^3 - (1-p)^5$ for all these codes, linear and non-linear. However, for non-linear $(n, q^k; q)$ codes C in general, we may sometimes have $P_{\text{ue}}(C,p) > (1-p)^{n-k} - (1-p)^n$. We will illustrate this with an explicit set of codes.

Example 2.15. Let $k \ge 4$ and $n = (q^k-1)/(q-1)$. Let C be the $(n, q^k; q)$ code containing the zero vector and all vectors of weight one. Then it is easy to see that

$$\begin{aligned} A_1(C) &= q(q-1)n/q^k, \\ A_2(C) &= (q-1)^2 n(n-1)/q^k, \\ A_i(C) &= 0 \text{ for } i \ge 3. \end{aligned}$$

Therefore

$$\begin{aligned} f(p) = P_{\text{ue}}(C,p) &= \frac{1}{q^k}\Big\{ q(q-1)n\frac{p}{q-1}(1-p)^{n-1} \\ &\quad + (q-1)^2 n(n-1)\Big(\frac{p}{q-1}\Big)^2 (1-p)^{n-2}\Big\} \\ &= \frac{1}{q^k}\Big\{ qnp(1-p)^{n-1} + n(n-1)p^2(1-p)^{n-2}\Big\}. \end{aligned}$$

Computing the derivative and evaluating it for $p = (q-1)/q$, we get

$$f'\Big(\frac{q-1}{q}\Big) = -\frac{1}{q^{k+n-1}(q-1)}(q^k-1)(q^{2k} - 2q^{k+1} - q^k + q^2).$$

Let $g(p) = (1-p)^{k-n} - (1-p)^n$. Then

$$g'\Big(\frac{q-1}{q}\Big) = -\frac{1}{q^{n-1}(q-1)}(q^{2k} - (q-1)kq^k + (q-1)^2).$$

Hence

$$\begin{aligned} &g'\Big(\frac{q-1}{q}\Big) - f'\Big(\frac{q-1}{q}\Big) \\ &= \frac{1}{q^{k+n-1}(q-1)}\Big((kq-k-2q)q^{2k} + q^2(q^k-1) + 2q^{k+1}\Big) > 0 \end{aligned}$$

if $kq - k - 2q \ge 0$. This is satisfied for $k \ge 4$ and all $q \ge 2$ (and also for $k = 3$ when $q \ge 3$). Hence

$$0 > g'\Big(\frac{q-1}{q}\Big) > f'\Big(\frac{q-1}{q}\Big)$$

for $k \ge 4$. Since $f\Big(\frac{q-1}{q}\Big) = g\Big(\frac{q-1}{q}\Big)$, this implies that $f(p) > g(p)$ when p is close to $(q-1)/q$, that is,

$$P_{\text{ue}}(C,p) > (1-p)^{k-n} - (1-p)^n$$

when $p \ge p_{q,k}$ for some $p_{q,k}$. For example, $p_{3,4} \approx 0.1687502258$. The first few values of $p_{2,k}$ are given in Table 2.15.

Table 2.9 The first few values of $p_{2,k}$ in Example 2.15.

k	4	5	6	7	8	9	10
$p_{2,k}$	0.28262	0.15184	0.08380	0.04677	0.02618	0.01463	0.00814

It is an open question in general to determine

$$\max\{P_{\rm ue}(C,p) \mid C \text{ is a } (n,M;q) \text{ code}\}$$

for given n, M, q, p.

The average results clearly give upper bounds on $P_{\rm ue}[n,k,p;q]$.

Theorem 2.52. *If K is any $[n,\kappa;q]$ code and $1 \le k \le \kappa$, then*

$$P_{\rm ue}[n,k,p;q] \le \frac{q^k-1}{q^\kappa-1} P_{\rm ue}(K,p)$$

for all $p \in \left[0, \frac{q-1}{q}\right]$. In particular, for all $n \ge \kappa \ge k \ge 1$ we have

$$P_{\rm ue}[n,k,p;q] \le \frac{q^k-1}{q^\kappa-1} P_{\rm ue}[n,\kappa,p;q].$$

Further, for all $n \ge k \ge 1$ we have

$$P_{\rm ue}[n,k,p;q] \le \frac{q^k-1}{q^n-1}\Big\{1-(1-p)^n\Big\},$$

and for all $n > k \ge 1$ we have

$$P_{\rm ue}[n,k,p;q] \le \frac{q^k-1}{q^n-q}\Big\{1+(q-1)\Big(1-\frac{q}{q-1}p\Big)^n - q(1-p)^n\Big\}.$$

Proof. The main result follows from Theorem 2.28, and the special cases by choosing $K = GF(q)^n$ and $K = \{(x_1, x_2, \ldots, x_n) \mid \sum_{i=1}^n x_i = 0\}$, respectively. □

There are sharper bounds in some cases.

Theorem 2.53. *For any integers $n \ge k \ge 1$, any $p \in \left[0, \frac{q-1}{q}\right]$, any $u \in (0,1]$, and any prime power q we have*

$$P_{\rm ue}[n,k,p;q] \le \Big(\frac{q^k-1}{q^n-1}\Big\{\Big((q-1)\Big(\frac{p}{q-1}\Big)^u + (1-p)^u\Big)^n - (1-p)^{un}\Big\}\Big)^{\frac{1}{u}}.$$

Proof. Let C be an $[n,k;q]$ code. Using the natural isomorphism between $GF(q)^n$ and $GF(q^n)$ we may consider C as a subset of $GF(q^n)$. For any non-zero $g \in GF(q^n)$, the set

$$gC = \{gx \mid x \in C\}$$

is again an $[n,k;q]$ code. For convenience we write

$$J = \Big(\frac{q^k-1}{q^n-1}\Big\{\Big((q-1)\Big(\frac{p}{q-1}\Big)^u + (1-p)^u\Big)^n - (1-p)^{un}\Big\}\Big)^{\frac{1}{u}},$$

the right-hand expression in the theorem. Define $\zeta(g)$ by

$$\zeta(g) = 1 \text{ if } P_{\mathrm{ue}}(gC,p) > J,$$
$$\zeta(g) = 0 \text{ if } P_{\mathrm{ue}}(gC,p) \le J.$$

Note that $\zeta(g) < \frac{P_{\mathrm{ue}}(gC,p)}{J}$ for all g. We will show that $\zeta(g) = 0$ for at least one g. First we show that

$$\zeta(g) < \sum_{x \in C\setminus\{0\}} \Big(\frac{\big(\frac{p}{q-1}\big)^{w_H(gx)}(1-p)^{n-w_H(gx)}}{J}\Big)^u. \tag{2.46}$$

If $\big(\frac{p}{q-1}\big)^{w_H(gx)}(1-p)^{n-w_H(gx)} > J$ for some $x \in C \setminus \{0\}$, then (2.46) is clearly satisfied. On the other hand, if $\big(\frac{p}{q-1}\big)^{w_H(gx)}(1-p)^{n-w_H(gx)} \le J$ for all $x \in C \setminus \{0\}$, then

$$\zeta(g) < \frac{P_{\mathrm{ue}}(gC,p)}{J} = \sum_{x \in C\setminus\{0\}} \Big(\frac{\big(\frac{p}{q-1}\big)^{w_H(gx)}(1-p)^{n-w_H(gx)}}{J}\Big)$$
$$\le \sum_{x \in C\setminus\{0\}} \Big(\frac{\big(\frac{p}{q-1}\big)^{w_H(gx)}(1-p)^{n-w_H(gx)}}{J}\Big)^u,$$

and (2.46) is again satisfied.

If $x \in C \setminus \{0\}$, then gx runs through $GF(q^n)$ when g does. Hence

$$\sum_{g \in GF(q^n)\setminus\{0\}} \zeta(g) < \sum_{g\neq 0} \sum_{x \in C\setminus\{0\}} \frac{1}{J^u}\Big(\frac{p}{q-1}\Big)^{uw_H(gx)}(1-p)^{u(n-w_H(gx))}$$
$$= \frac{1}{J^u} \sum_{x \in C\setminus\{0\}} \sum_{g\neq 0} \Big(\frac{p}{q-1}\Big)^{uw_H(gx)}(1-p)^{u(n-w_H(gx))}$$
$$= \frac{1}{J^u} \sum_{x \in C\setminus\{0\}} \Big\{\Big((q-1)\Big(\frac{p}{q-1}\Big)^u + (1-p)^u\Big)^n - (1-p)^{un}\Big\}$$
$$= \frac{q^k-1}{J^u}\Big\{\Big((q-1)\Big(\frac{p}{q-1}\Big)^u + (1-p)^u\Big)^n - (1-p)^{un}\Big\}$$
$$= q^n - 1.$$

Therefore $\zeta(g) = 0$ for at least one $g \neq 0$. □

Lemma 2.8. *If $R = \frac{k}{n}$ and*

$$u = \frac{\log_q \frac{(q-1)(1-\rho(R))}{\rho(R)}}{\log_q \frac{(q-1)(1-p)}{p}},$$

then

$$\Big(\frac{q^k-1}{q^n-1}\Big\{\Big((q-1)\Big(\frac{p}{q-1}\Big)^u+(1-p)^u\Big)^n-(1-p)^{un}\Big\}\Big)^{\frac{1}{u}} \le \Big(\frac{p}{q-1}\Big)^{n\rho(R)}(1-p)^{n-n\rho(R)}$$

Proof. For this value of u we have

$$\frac{(q-1)(1-\rho(R))}{\rho(R)} = \Big(\frac{(q-1)(1-p)}{p}\Big)^u$$

and so $(q-1)\Big(\frac{p}{q-1}\Big)^u + (1-p)^u = \frac{(q-1)\left(\frac{p}{q-1}\right)^u}{\rho(R)}$. Further,

$$q^{R-1} = q^{-H_q(\rho(R))} = \Big(\frac{\rho(R)}{q-1}\Big)^{\rho(R)}(1-\rho(R))^{1-\rho(R)}.$$

Hence

$$\begin{aligned}
J &\le \Big\{q^{k-n}\Big((q-1)\Big(\frac{p}{q-1}\Big)^u+(1-p)^u\Big)^n\Big\}^{\frac{1}{u}} \\
&= \Big\{q^{R-1}\Big((q-1)\Big(\frac{p}{q-1}\Big)^u+(1-p)^u\Big)\Big\}^{\frac{n}{u}} \\
&= \Big\{\Big(\frac{\rho(R)}{q-1}\Big)^{\rho(R)}(1-\rho(R))^{1-\rho(R)}\frac{(q-1)\left(\frac{p}{q-1}\right)^u}{\rho(R)}\Big\}^{\frac{n}{u}} \\
&= \Big\{\Big(\frac{p}{q-1}\Big)^u\Big(\frac{(q-1)(1-\rho(R))}{\rho(R)}\Big)^{1-\rho(R)}\Big\}^{\frac{n}{u}} \\
&= \Big\{\Big(\frac{p}{q-1}\Big)^u\Big(\frac{(q-1)(1-p)}{p}\Big)^{u(1-\rho(R))}\Big\}^{\frac{n}{u}} \\
&= \Big\{\Big(\frac{p}{q-1}\Big)^{\rho(R)}(1-p)^{1-\rho(R)}\Big\}^n.
\end{aligned}$$

□

Combining Theorem 2.53 and Lemma 2.8 we get the following theorem.

Theorem 2.54. *If $p \in \left[0, \frac{1}{q}\right]$ and $R = \frac{k}{n} \le C(p) = 1 - H_q(p)$ (that is, $\rho(R) \ge p$), then*

$$P_{\text{ue}}[n,k,p;q] \le \Big(\frac{p}{q-1}\Big)^{n\rho(R)}(1-p)^{n-n\rho(R)}.$$

2.6.3 *Asymptotic bounds*

It turns out that $P_{\rm ue}(n, M, p; q)$ decreases exponentially with n. Therefore, define

$$\pi_{\rm ue}(n, R, p; q) = -\frac{\log_q P_{\rm ue}(n, \lceil q^{Rn} \rceil, p; q)}{n},$$

$$\overline{\pi}_{\rm ue}(R, p; q) = \limsup_{n\to\infty} \pi_{\rm ue}(n, R, p; q),$$

$$\underline{\pi}_{\rm ue}(R, p; q) = \liminf_{n\to\infty} \pi_{\rm ue}(n, R, p; q),$$

and

$$\pi_{\rm ue}(R, p; q) = \lim_{n\to\infty} \pi_{\rm ue}(n, R, p; q)$$

when the limit exists. A lower (resp. upper) bound on $P_{\rm ue}(n, M; q)$ will give an upper (resp. lower) bound on $\pi_{\rm ue}(n, R, p; q)$.

From Theorems 2.48 we get the following bound.

Theorem 2.55. *For* $0 \le R_1 \le R_2 \le 1$ *we have*

$$\overline{\pi}_{\rm ue}(R_2, p; q) \le -\delta(R_1)\log_q\Big(\frac{p}{q-1}\Big) - (1-\delta(R_1))\log_q(1-p) + R_1 - R_2.$$

From Theorem 2.44 we can get upper bounds on $\overline{\pi}_{\rm ue}(R, p; q)$. The sum contains $\lfloor \log_q(M) \rfloor + 1$ terms. For the term for $i = \omega n$ we get

$$-\frac{1}{n}\log_q\Big(\binom{n}{\omega n}(Mq^{\omega n - n} - 1)\Big(\frac{p}{q-1}\Big)^{\omega n}\Big(1 - \frac{qp}{q-1}\Big)^{n-\omega n}\Big) \sim f(\omega)$$

where

$$f(\omega) = \omega\log_q(\omega) + (1-\omega)\log_q(1-\omega) - (R+\omega-1)$$
$$-\omega\log_q\Big(\frac{p}{q-1}\Big) - (1-\omega)\log_q\Big(\frac{qp}{q-1}\Big).$$

The function $f(\omega)$ takes its minimum for $\omega = \frac{qp}{q-1}$ and the corresponding minimum is $1 - R$. The smallest value for ω with i in the summation range is $\omega = 1 - R$. Hence, if $\frac{qp}{q-1} < 1 - R$, then $f(\omega) \ge f(1-R)$. This gives the following theorem.

Theorem 2.56. *(i) For* $1 - \frac{qp}{q-1} \le R \le 1$ *we have*

$$\overline{\pi}_{\rm ue}(R, p; q) \le 1 - R.$$

(ii) For $0 \le R \le 1 - \frac{qp}{q-1}$ *we have*

$$\overline{\pi}_{\rm ue}(R, p; q) \le R\Big\{\log_q(R) - \log_q\Big(1 - \frac{qp}{q-1}\Big)\Big\}$$
$$+(1-R)\Big\{\log_q(1-R) - \log_q\Big(\frac{qp}{q-1}\Big)\Big\}.$$

Theorem 2.57. *For $0 \le R \le C(p)$ we have*

$$\underline{\pi}_{\rm ue}(R,p;q) \ge -\rho(R)\log_q\Big(\frac{p}{q-1}\Big) - (1-\rho(R))\log_q(1-p).$$

For $0 \le R \le 1$ we have

$$\underline{\pi}_{\rm ue}(R,p;q) \ge 1-R.$$

Proof. The first bound follows directly from Theorem 2.54. From Theorem 2.43 we get

$$P_{\rm ue}(n,q^k,p;q) \le P_{\rm ue}[n,k,p;q] \le q^{k-n}$$

and so

$$\frac{\log_q P_{\rm ue}(n,\lceil q^{Rn}\rceil,p;q)}{n} \le R-1$$

for $0 \le R \le 1$. This proves the second bound. The two bounds coincide when $R = C(p)$; the first bound is stronger when $R < C(p)$. □

Combining Theorems 2.56 and 2.57 we get the following theorem.

Theorem 2.58. *For $1 - \frac{qp}{q-1} \le R \le 1$ we have*

$$\pi_{\rm ue}(R,p;q) = 1-R.$$

Any upper bound on $\delta(R)$ will give an upper bound on $\overline{\pi}_{\rm ue}(R,p)$. The best upper bounds on $\delta(R)$ known is the LP-bound (see e.g. MacWilliams and Sloane (1977, Chapter 17)). Using any upper bound on $\delta(R)$, the first of the bounds of Theorem 2.55 is better for small values of p, the other for larger values of p.

Results that are sometimes stronger than Theorem 2.56 are known. They are given for the binary case and we list those here, but omit the proofs. To formulate the results, some additional notation is needed. Let

$$\delta_{\rm LP1}(R) = \frac{1}{2} - \sqrt{H_2^{-1}(R)(1-H_2^{-1}(R))}.$$

Theorem 2.59. *For $0 \le R \le R_{\rm LP1}(p)$ we have*

$$\overline{\pi}_{\rm ue}(R,p;2) \le 1 - R - H_2(\delta_{\rm LP1}(R)) + T(\delta_{\rm LP1}(R),p).$$

For $0 \le R \le R_{\rm LP}(p)$ we have

$$\overline{\pi}_{\rm ue}(R,p;2) \le 1 - R - H_2(\delta_{\rm LP}(R)) + T(\delta_{\rm LP}(R),p).$$

For $\min\{R_{\rm LP1}(p), R_{\rm LP}(p)\} < R \le 1$ we have

$$\pi_{\rm ue}(R,p;2) = 1-R.$$

2.7 Optimal codes

2.7.1 *The dual of an optimal code*

Theorem 2.60. *If C is an $[n, k; q]$ code which is optimal for p, then $C^{\perp}$ is optimal for $\frac{q-1-qp}{q-qp}$. Moreover,*

$$P_{\rm ue}[n, k, p; q] = q^k(1-p)^n P_{\rm ue}\left[n, n-k, \frac{q-1-qp}{q-qp}\right] + q^{k-n} - (1-p)^n.$$

Proof. Theorem 2.7 implies that if $P_{\rm ue}(C_1, p) \le P_{\rm ue}(C_2, p)$, then

$$P_{\rm ue}\left(C_1^{\perp}, \frac{q-1-qp}{q-qp}\right) \le P_{\rm ue}\left(C_q^{\perp}, \frac{q-1-qp}{q-qp}\right),$$

and the theorem follows. □

Since $\frac{q-1-qp}{q-qp}$ runs through $\left[0, \frac{q-1}{q}\right]$ when p does, we have the following corollary.

Corollary 2.12. *If C is a code which is optimal for all $p \in \left[0, \frac{q-1}{q}\right]$, then so is $C^{\perp}$.*

2.7.2 *Copies of the simplex code*

Let k and s by any positive integers. In Subsection 1.2.2, we described the simplex codes and the corresponding generator matrices Γ_k. Let C be the $\left[\frac{s(q^k-1)}{q-1}, k; q\right]$ code generated by the matrix obtained by concatenating s copies of Γ_k. This code has minimum distance $d = sq^{k-1}$. From Theorems 2.42 and 2.49 we get

$$P_{\rm ue}(C, p) = (q^k - 1)\left(\frac{p}{q-1}\right)^d (1-p)^{n-d} = P_{\rm ue}\left[\frac{s(q^k-1)}{q-1}, k, p; q\right].$$

Hence the code C is optimal. In particular, for $s = 1$ we get the simplex code whose dual is the Hamming code. Hence the Hamming codes are also optimal.

In the next chapter (in Section 3.2) we describe the optimal binary codes of dimension four or less.

2.8 New codes from old

As illustrated in the previous chapter, there are a number of ways to construct new codes from one or two old ones. For many of these constructions, the new code may not be good even if the old is/are. We shall give a number of examples which illustrate this. The *-operation is atypical in this respect. It gives a systematic way to make long good codes and we discuss this construction first.

2.8.1 The *-operation

Theorem 2.61. *Let C be linear $[n,k;q]$ code.*
(i) If C is good, then C^* is good.
(ii) If C is proper, then C^* is proper.

Proof. Since $\left(\frac{p}{q-1}\right)^{q^{k-1}}(1-p)^{\frac{q^{k-1}-1}{q-1}}$ is increasing on $\left[0, \frac{q-1}{q}\right]$ and

$$P_{\rm ue}(C^*,p) = \left(\frac{p}{q-1}\right)^{q^{k-1}}(1-p)^{\frac{q^{k-1}-1}{q-1}} P_{\rm ue}(C,p),$$

(ii) follows. Further,

$$\left(\frac{p}{q-1}\right)^{q^{k-1}}(1-p)^{\frac{q^{k-1}-1}{q-1}} \le \left(\frac{1}{q}\right)^{\frac{q^k-1}{q-1}},$$

and so *(i)* follows. □

Theorem 2.61 shows that we can make longer good codes from shorter ones. However, there is a stronger result which shows that starting from any code and using the *-operation repeatedly, we will eventually get a good code, and even a proper code.

Theorem 2.62. *If C is an $[n,k,d;q]$ code and $r \ge \max\{0,(q-1)n-qd\}$, then C^{r*} is proper.*

Proof. If $d \ge \frac{q-1}{q}n$, then C is proper by Theorem 2.15. Consider $d < \frac{q-1}{q}n$. Then $r > 0$ and C^{r*} is an $\left[n + r\frac{q^k-1}{q-1}, k, d + rq^{k-1}; q\right]$ code. If

$$d + rq^{k-1} \ge \frac{q-1}{q}\left\{n + r\frac{q^k-1}{q-1}\right\},$$

that is, $r \ge (q-1)n - qd$, then C^{r*} is proper by Theorem 2.15. □

If $r \ge \max\{0,(q-1)n-q\}$, then the condition on r is satisfied for all $d \ge 1$. Hence we get the following corollary.

Corollary 2.13. *If C is an $[n,k;q]$ code and $r \ge (q-1)n-q$, then C^{r*} is proper.*

We can get the weaker conclusion that C^{r*} is good under a condition on r that is sometimes weaker.

Theorem 2.63. *If C is an $[n, k; q]$ code and $r \geq (q-1)(n-k)$, then C^{r*} is good.*

Proof. By Theorem 2.51, $P_{\rm ue}(C,p) \leq (1-p)^{n-k}\Big\{1-(1-p)^k\Big\}$. Hence,

$$\begin{aligned}
P_{\rm ue}(C^{r*},p) &= \Big\{\Big(\frac{p}{q-1}\Big)^{q^{k-1}}(1-p)^{\frac{q^{k-1}-1}{q-1}}\Big\}^r P_{\rm ue}(C,p)\\
&\leq \Big\{\Big(\frac{p}{q-1}\Big)^{q^{k-1}}(1-p)^{\frac{q^{k-1}-1}{q-1}}\Big\}^r (1-p)^{n-k}\Big\{1-(1-p)^k\Big\}\\
&= \Big(\frac{p}{q-1}\Big)^{rq^{k-1}}(1-p)^{r\frac{q^{k-1}-1}{q-1}+(n-k)}\Big\{1-(1-p)^k\Big\}.
\end{aligned}$$

Clearly, $1-(1-p)^k$ is increasing on $[0,(q-1)/q]$. If $r \geq (q-1)(n-k)$, then

$$rq^{k-1} \geq \frac{q-1}{q}\Big\{rq^{k-1} + r\frac{q^{k-1}-1}{q-1} + (n-k)\Big\},$$

and $p^{rq^{k-1}}(1-p)^{r\frac{q^{k-1}-1}{q-1}+(n-k)}$ is also increasing on $[0,(q-1)/q]$. Hence we get

$$\begin{aligned}
P_{\rm ue}(C^{r*},p) &\leq \Big(\frac{1}{q}\Big)^{rq^{k-1}+r\frac{q^{k-1}-1}{q-1}+(n-k)}\Big\{1-\Big(\frac{1}{q}\Big)^k\Big\}\\
&= \frac{q^k-1}{q^{n+r\frac{q^k-1}{q-1}}} = P_{\rm ue}\Big(C^{r*},\frac{q-1}{q}\Big).
\end{aligned}$$

□

In Remark 2.7 following Theorem 2.51 we gave an $[n, k; q]$ code C for which $P_{\rm ue}(C,p) = (1-p)^{n-k}\Big\{1-(1-p)^k\Big\}$. It is easy to check that for this code and $r = (q-1)(n-k)-1$ we get

$$\frac{dP_{\rm ue}(C^{r*},p)}{dp}\Big|_{p=(q-1)/q} = q^{-\left(r\frac{q^k-1}{q-1}+n\right)}\Big\{k-\frac{q^k-1}{q-1}\Big\} < 0.$$

Hence, C^{r*} is bad. This shows that the bound on r in Theorem 2.63 cannot be improved in general. It is an open question if the bound on r in Theorem 2.62 can be sharpened in general.

As a final remark, we note that if $(q-1)(n-k) > (q-1)n - qd > 0$, that is,

$$\frac{q-1}{q}k < d < \frac{q-1}{q}n,$$

then Theorem 2.63 is weaker than Theorem 2.62.

From Corollary 2.13, we get the following theorem.

Theorem 2.64. *For any q, k, and $n \geq \frac{(q^k-1-q)(q^k-1)}{q-1}$ there exists a proper $[n,k;q]$ code.*

Proof. Let $n = r\frac{q^k-1}{q-1} + \nu$ where $0 \leq \nu \leq \frac{q^k-1}{q-1} - 1$ and $r \geq q^k - 1 - q$. First, consider $\nu \geq k$. Let C_0 some $[\nu,k;q]$ code. Since

$$r \geq q^k - 1 - q = (q-1)\Big(\frac{q^k-1}{q-1} - 1\Big) - q \geq (q-1)\nu - q,$$

Corollary 2.13 implies that C_0^{r*} is a proper code of length n.

Next consider $\nu < k$. Let C_1 be the $\frac{q^k-1}{q-1} + \nu$ code generated by the matrix

$$\left(\Gamma_k \,\middle|\, \begin{matrix} I_\nu \\ O \end{matrix}\right),$$

that is, Γ_k concatenated by ν distinct unit vectors. The weight distribution of C_1 is easily seen to be:

$$\begin{aligned} A_0 &= 1, \\ A_{(q^k-1)/(q-1)} &= q^{k-\nu} - 1, \\ A_{(q^k-1)/(q-1)+i} &= \binom{\nu}{i}(q-1)^i q^{k-\nu} \text{ for } 1 \leq i \leq \nu. \end{aligned}$$

Hence

$$\begin{aligned} P_{\mathrm{ue}}(C_1,p) &= (q^{k-\nu}-1)\Big(\frac{p}{q-1}\Big)^{\frac{q^k-1}{q-1}}(1-p)^{\frac{q^{k-1}-1}{q-1}+\nu} \\ &\quad + q^{k-\nu}\sum_{i=1}^{\nu}\binom{\nu}{i}(q-1)^i\Big(\frac{p}{q-1}\Big)^{\frac{q^k-1}{q-1}+i}(1-p)^{\frac{q^{k-1}-1}{q-1}+\nu-i} \\ &= \Big(\frac{p}{q-1}\Big)^{\frac{q^k-1}{q-1}}(1-p)^{\frac{q^{k-1}-1}{q-1}}\Big\{q^{k-\nu} - (1-p)^\nu\Big\}. \end{aligned}$$

Since both $\left(\frac{p}{q-1}\right)^{\frac{q^k-1}{q-1}}(1-p)^{\frac{q^{k-1}-1}{q-1}}$ and $q^{k-\nu} - (1-p)^\nu$ are increasing on $[0,(q-1)/q]$, C_1 is proper. Hence $C_1^{(r-1)*}$ is a proper code of length n. □

We believe that proper $(n,M;q)$ codes exist for all q, n and M, but this is an open question.

2.8.2 Shortened codes

Consider a systematic $[n, k; q]$ code C and let C^s be the shortened $[n-1, k-1]$ code:

$$C^s = \Big\{ \mathbf{x} \in GF(q)^{n-1} \;\Big|\; (0|\mathbf{x}) \in C \Big\}.$$

Again the shortened code may be good or bad, whether the original code is good or not, as shown by the examples in Figure 2.1.

	C^s proper	C^s bad
C proper	$\begin{pmatrix} 100 \\ 010 \\ 001 \end{pmatrix}$	$\begin{pmatrix} 1001 \\ 0100 \\ 0010 \end{pmatrix}$
$A_C(z)$:	$1 + 3z + 3z^2 + z^3$	$1 + 2z + 2z^2 + 2z^3 + z^4$
$A_{C^s}(z)$:	$1 + 2z + z^2$	$1 + 2z + z^2$
$\Delta_C(z)$:	z	$-z^2 + z^3 + z^4$
C bad	$\begin{pmatrix} 1000000 \\ 0101011 \\ 0010111 \end{pmatrix}$	$\begin{pmatrix} 100111 \\ 010000 \\ 001000 \end{pmatrix}$
$A_C(z)$:	$1 + z + 3z^4 + 3z^5$	$1 + 2z + z^2 + z^4 + 2z^5 + z^6$
$A_{C^s}(z)$:	$1 + 3z^4$	$1 + 2z + z^2$
$\Delta_C(z)$:	z	$-2z^2 - z^3 + 2z^4 + 2z^5$

Fig. 2.1 Examples of some shortened binary codes.

We note that

$$P_{\text{ue}}\Big(C^s, \frac{q-1}{q}\Big) = \frac{q^{k-1}-1}{q^{n-1}} < \frac{q^k - 1}{q^n} = P_{\text{ue}}\Big(C, \frac{q-1}{q}\Big).$$

It is natural to ask if $P_{\text{ue}}(C^s, p) \le P_{\text{ue}}(C, p)$ for all $p \in \left[0, \frac{q-1}{q}\right]$. In terms of the weight distribution this can be written as

$$\Delta_C(z) \stackrel{\text{def}}{=} \Big\{ A_C(z) - 1 \Big\} - (1+z) \Big\{ A_{C^s}(z) - 1 \Big\} \ge 0$$

for all $z \in [0, 1]$. The examples in Figure 2.1 show that this may or may not be true. For these examples, $q = 2$, the codes are linear and they are given by their generator matrices. Clearly $z \geq 0$ for $z \in [0,1]$. But, $-z^2+z^3+z^4<0$ for $z \in [0, \frac{\sqrt{5}-1}{2})$, and $-2z^2-z^3+2z^4+2z^5<0$ for $z \in [0, 0.87555]$.

2.8.3 *Product codes*

Let C_1 be an $[n_1, k_1, d_1; q]$ code and C_2 an $[n_2, k_2, d_2; q]$ code. Let C be the product code. This has length $n = n_1 n_2$, dimension $k = k_1 k_2$, and minimum distance $d = d_1 d_2$. There is in general no simple expression for $A_C(z)$ in terms of $A_{C_1}(z)$ and $A_{C_2}(z)$.

Theorem 2.65. *Let d_1 and d_2 be fixed. The product code C of an $[n_1, k_1, d_1; q]$ code C_1 and an $[n_2, k_2, d_2; q]$ code C_2 where $n_1 > k_1$ and $n_1 \geq n_2 > k_2$ is bad if n_1 is sufficiently large.*

Proof. First we note that

$$n-k = n_1 n_2 - k_1 k_2 \geq n_1 n_2 - (n_1-1)(n_2-1) = n_1 + n_2 - 1 \geq n_1.$$

Rough estimates give

$$\begin{aligned}
\frac{P_{\text{ue}}(C, \frac{d}{n})}{P_{\text{ue}}(C, \frac{q-1}{q})} &\geq q^{n-k}\Big(\frac{d}{(q-1)n}\Big)^d\Big(1-\frac{d}{n}\Big)^{n-d} \\
&\geq q^{n_1}\Big(\frac{d}{(q-1)n}\Big)^d\Big(1-\frac{d}{n}\Big)^{n} \\
&= q^{n_1}\Big(\frac{d}{(q-1)}\Big)^d\frac{1}{n^d}\Big\{\Big(1-\frac{d}{n}\Big)^{n/d}\Big\}^d \\
&\geq q^{n_1}\Big(\frac{d}{(q-1)}\Big)^d\frac{1}{(n_1 n_2)^d}\Big(\frac{1}{4}\Big)^d
\end{aligned}$$

if $n \geq 2d$. Since q^{n_1} is exponential in n_1 whereas $(n_1 n_2)^d$ is polynomial, the theorem follows. □

2.8.4 *Repeated codes*

Let C be an $(n, M, d; q)$ code and let $r \geq 1$. From the definition of C^r, we see that

$$A_{C^r}(z) = \Big\{A_C(z)\Big\}^r.$$

In particular,

$$A_d(C^r) = rA_d(C).$$

Theorem 2.66. *If C is an $(n, M, d; q)$ code where $1 < M < q^n$, and $0 < p \le 1 - \frac{M^{1/n}}{q}$, then*

$$\frac{P_{\text{ue}}(C^r, p)}{P_{\text{ue}}(C^r, (q-1)/q)} \to \infty \text{ when } r \to \infty.$$

Proof. The condition $0 < p \le 1 - \frac{M^{1/n}}{q}$ is equivalent to $(1-p)^n \ge \frac{M}{q^n}$. Hence

$$\begin{aligned} P_{\text{ue}}(C^r, p) &\ge r A_d(C) \Big(\frac{p}{(q-1)(1-p)}\Big)^d (1-p)^{rn} \\ &> r A_d(C) \Big(\frac{p}{(q-1)(1-p)}\Big)^d \cdot \frac{M^{rn}-1}{q^{rn}} \\ &= r A_d(C) \Big(\frac{p}{(q-1)(1-p)}\Big)^d P_{\text{ue}}\Big(C^r, \frac{q-1}{q}\Big) \end{aligned}$$

and so

$$\frac{P_{\text{ue}}(C^r, p)}{P_{\text{ue}}(C^r, (q-1)/q)} > r A_d(C) \Big(\frac{p}{(q-1)(1-p)}\Big)^d \to \infty$$

when $r \to \infty$. □

Corollary 2.14. *If C is an $(n, M, d; q)$ and r is sufficiently large, then C^r is bad.*

A careful analysis shows that C^r is bad for all $r \ge 5$.

2.9 Probability of having received the correct code word

If C is an $(n, M; q)$ and a code word from C is transmitted over a symmetric channel with symbol error probability p, then the probability that the code word is received without errors is $(1-p)^n$. The probability that another code word is received is, by definition, $P_{\text{ue}}(C, p)$. Hence the probability that we receive *some* code word is

$$(1-p)^n + P_{\text{ue}}(C, p),$$

and the probability that we have received the correct code word under the condition that we have received some code word is

$$P_{\text{corr}}(C, p) = \frac{(1-p)^n}{(1-p)^n + P_{\text{ue}}(C, p)}.$$

Theorem 2.67. *Let C be an $(n, M; q)$ code. The function $P_{\text{corr}}(C, p)$ is strictly decreasing on $[0, (q-1)/q]$,*

$$P_{\text{corr}}(C, 0) = 1 \quad \text{and} \quad P_{\text{corr}}(C, (q-1)/q) = 1/M.$$

Proof. We have

$$\frac{1}{P_{\rm corr}(C,p)} = 1 + \frac{P_{\rm ue}(C,p)}{(1-p)^n} = \sum_{i=0}^{n} A_i(C)\Big(\frac{p}{(q-1)(1-p)}\Big)^i.$$

Since $\frac{p}{(q-1)(1-p)}$ is strictly increasing on $[0,(q-1)/q]$, $1/P_{\rm corr}(C,p)$ is strictly increasing also, and so $P_{\rm corr}(C,p)$ is strictly decreasing. □

Any lower (resp. upper) bound on $P_{\rm ue}(C,p)$ gives an upper (resp. lower) bound on $P_{\rm corr}(C,p)$. For example, from Theorem 2.51 we get the following result.

Theorem 2.68. *Let C be an $[n,k;q]$ code. Then $P_{\rm corr}(C,p) \geq (1-p)^k$.*

By Theorem 2.8, the average value of $P_{\rm ue}(C,p)$ over all $(n,M;q)$ codes is $\frac{M-1}{q^n-1}\Big\{1-(1-p)^n\Big\}$. Hence, we get the following result.

Theorem 2.69. *The average value of $1/P_{\rm corr}(C,p)$ over all $(n,M;q)$ codes is*

$$\frac{q^n-M}{q^n-1} + \frac{M-1}{q^n-1}\frac{1}{(1-p)^n}.$$

From Theorem 2.69 we also get an asymptotic result.

Theorem 2.70. *Let $C_{R,n}$ be an average $(n,q^{Rn};q)$ code of rate R. Then*

$$\lim_{n\to\infty} P_{\rm corr}(C_{R,n},p) = \begin{cases} 1 \text{ for } p < 1-q^{R-1}, \\ 0 \text{ for } p > 1-q^{R-1}. \end{cases}$$

Proof. We see that

$$\frac{q^n-q^{Rn}}{q^n-1} \sim 1,$$

and

$$\frac{M-1}{q^n-1}\frac{1}{(1-p)^n} \sim \Big(\frac{q^{R-1}}{1-p}\Big)^n$$

when $n\to\infty$. If $q^{R-1} < 1-p$, then $1/P_{\rm corr}(C_{R,n},p)\to 1$ by Theorem 2.69. Similarly, $1/P_{\rm corr}(C_{R,n},p)\to\infty$ if $q^{R-1} > 1-p$. □

2.10 Combined correction and detection

2.10.1 *Using a single code for correction and detection*

Let C be an $(n, M; q)$ code. Let $P_{\text{ue}}^{(t)}(C, p)$ be the probability of having an undetected error after the correction of t or less errors (after transmission over the q-ary symmetric channel). A more detailed discussion of $P_{\text{ue}}^{(t)}(C, p)$ is outside the scope of this book, but we briefly mention some basic results.

In 1.6.2 we showed that the probability of undetected error after using the (n, M, d) code C to correct t or less errors, where $2t + 1 \leq d$, is given by

$$P_{\text{ue}}^{(t)}(C, p) = \sum_{j=d-t}^{n} A_{t,j} \left(\frac{p}{q-1}\right)^j (1-p)^{n-j} \tag{2.47}$$

where

$$A_{t,j} = \sum_{i=j-t}^{j+t} A_i N_t(i, j)$$

and $N_t(i, j)$ is given by Theorem 1.21.

Any question considered for $P_{\text{ue}}(C, p)$ may be considered for $P_{\text{ue}}^{(t)}(C, p)$ for $t > 0$, e.g. bounds, means, etc. In particular, C is called *t-proper* if $P_{\text{ue}}^{(t)}(C, p)$ is monotonous and *t-good* if

$$P_{\text{ue}}^{(t)}(C, p) \leq P_{\text{ue}}^{(t)}\left(C, \frac{q-1}{q}\right) \tag{2.48}$$

for all $p \in \left[0, \frac{q-1}{q}\right]$.

In Figure 2.2 we give examples of generator matrices of two $[12, 4; 2]$ codes which show that a code C may be proper without being 1-proper, or good without being 1-good.

A simple upper bound

Let C be an $[n, k, d; q]$ code where $d \geq 2t + 1$. When $\mathbf{0}$ is sent and $\mathbf{y}$ is received, then we have an undetected error, after correcting of t or less errors, if and only of $\mathbf{y} \in \bigcup_{\mathbf{x} \in C \setminus \{\mathbf{0}\}} S_t(\mathbf{x})$. By definition, the spheres $S_t(\mathbf{x})$ are disjoint and the union is a subset of $GF(q)^n$. Hence $A_{t,j} \leq \binom{n}{j}(q-1)^j$ and we get the following upper bound.

Theorem 2.71. *Let C be an $[n, k, d; q]$ code with $d \geq 2t + 1$. Then*

$$P_{\text{ue}}^{(t)}(C, p) \leq \sum_{j=d-t}^{n} \binom{n}{j} p^j (1-p)^{n-j}.$$

C is proper is not 1-proper	$\begin{pmatrix} 100011110000 \\ 010001101111 \\ 001000010001 \\ 000100010011 \end{pmatrix}$
$A_C^{(0)}(z)$:	$2z^3 + z^4 + z^5 + 2z^6 + 4z^7 + 4z^8 + z^9$
$A_C^{(1)}(z)$:	$6z^2 + 6z^3 + 24z^4 + 21z^5 + 37z^6$ $+48z^7 + 33z^8 + 17z^9 + 3z^{10}$
C is good is not 1-good	$\begin{pmatrix} 100011100101 \\ 010000011001 \\ 001000011111 \\ 000100011011 \end{pmatrix}$
$A_C^{(0)}(z)$:	$2z^3 + 2z^4 + z^5 + 2z^6 + 2z^7 + 3z^8 + 3z^9$
$A_C^{(1)}(z)$:	$6z^2 + 10z^3 + 25z^4 + 29z^5 + 23z^6$ $+38z^7 + 40z^8 + 15z^9 + 9z^{10}$

Fig. 2.2 Examples of two $[12, 4; 2]$ codes, given by generator matrices.

The ratio $P_{\text{ue}}^{(t)}(C,p)/P_{\text{ue}}(C,p)$

We now consider how the probability of undetected error increases when some of the power of the code is used for error correction.

For $p = \frac{q-1}{q}$ we have

$$\frac{P_{\text{ue}}^{(t)}\left(C, \frac{q-1}{q}\right)}{P_{\text{ue}}\left(C, \frac{q-1}{q}\right)} = \sum_{j=0}^{t} \binom{n}{j} (q-1)^j.$$

In general

$$(1-p)^{-n} P_{\text{ue}}(C,p) = A_d z^d + A_{d+1} z^{d+1} + \cdots$$

and

$$(1-p)^{-n} P_{\text{ue}}^{(t)}(C,p) = A_{t,d-t} z^{d-t} + A_{t,d-t+1} z^{d-t+1} + \cdots$$

where $z = p/((q-1)(1-p))$ and

$$A_{t,d-t} = \binom{d}{d-t} A_d$$

$$A_{t,d-t+1} = \binom{d}{d-t+1} \Big\{1 + (d-t+1)(q-2)\Big\} A_d + \binom{d+1}{d-t+1} A_{d+1}.$$

Hence,

$$\frac{P_{\text{ue}}^{(t)}(C,p)}{P_{\text{ue}}(C,p)} = \binom{d}{t} z^{-t}\Big\{1 - t\Big(\frac{A_{d+1}/A_d + 1}{d-t+1} + (q-2)\Big)z + \cdots\Big\}.$$

More terms can be added, but they soon become quite complicated.

Average

One can find average results for $P_{\text{ue}}^{(t)}(p)$ similar to those found for $P_{\text{ue}}(p)$ in Section 2.4. The proof is also similar and we omit the details. The result can be expressed in various ways, and we list some.

Theorem 2.72. *Let $\mathcal{C}$ be a set of codes of length n and minimum distance at least $d = 2t+1$. Then*

$$\begin{aligned} P_{\text{ue}}^{(t)}(\mathcal{C},p) &= \sum_{w_H(\mathbf{x})\ge 2t+1} \alpha(\mathbf{x}) \sum_{\mathbf{y}\in S_t(\mathbf{x})} \Big(\frac{p}{q-1}\Big)^{w_H(\mathbf{y})} (1-p)^{n-w_H(\mathbf{y})} \\ &= \sum_{i=2t+1}^{n} A_i(\mathcal{C}) \sum_{j=i-t}^{i+t} N_t(i,j)\Big(\frac{p}{q-1}\Big)^j (1-p)^{n-j} \\ &= \sum_{w_H(\mathbf{y})\ge t+1} \Big(\frac{p}{q-1}\Big)^{w_H(\mathbf{y})} (1-p)^{n-w_H(\mathbf{y})} \sum_{\mathbf{x}\in S_t(\mathbf{y})} \alpha(\mathbf{x}). \end{aligned}$$

We consider one example.

Theorem 2.73. *For the set $\mathcal{C} = \mathcal{SYSL}(n,k,2t+1)$ we have*

$$P_{\text{ue}}^{(t)}(\mathcal{C},p) \le \frac{\sum_{i=0}^{t}\binom{n}{i}}{q^{n-k} - \sum_{i=0}^{d-2}\binom{n-1}{i}} \Big\{1 - \sum_{i=0}^{t}\binom{n}{i}\Big(\frac{p}{q-1}\Big)^i (1-p)^{n-i}\Big\}.$$

Proof. Let

$$\beta = \frac{1}{q^{n-k} - \sum_{i=0}^{d-2}\binom{n-1}{i}(q-1)^i}.$$

By (2.27) we have $\alpha(\mathbf{x}) \le \beta$ for all $\mathbf{x}$ of weight at least $2t+1$. Hence

$$\begin{aligned} P_{\text{ue}}^{(t)}(\mathcal{C},p) &\le \beta \sum_{w_H(\mathbf{y})\ge t+1} \Big(\frac{p}{q-1}\Big)^{w_H(\mathbf{y})} (1-p)^{n-w_H(\mathbf{y})} \#S_t(\mathbf{y}) \\ &= \beta\Big\{1 - \sum_{i=0}^{t}\binom{n}{i}\Big(\frac{p}{q-1}\Big)^i (1-p)^{n-i}\Big\} \sum_{i=0}^{t}\binom{n}{i}. \end{aligned}$$

□

2.10.2 *Concatenated codes for error correction and detection*

Two types of concatenated codes for combined error correction and detection have been studied.

$m = 1$

Let the outer code D be a $[k, l; q]$ code and the inner code C an $[n, k; q]$ systematic code. Let $u(\mathbf{x}) \in GF(q)^{n-k}$ be the check bits corresponding to the information vector $\mathbf{x} \in GF(q)^k$, and let

$$E = \{(\mathbf{x}|u(\mathbf{x})) \mid \mathbf{x} \in D\}.$$

Then E is an $[n, l; q]$ code. Suppose without loss of generality that the all-zero code word is sent. At the receiving end we first use complete maximum likelihood decoding for the code C, that is we decode into the closest code word of $(\mathbf{x}|u(\mathbf{x})) \in C$, and if there is more than one closest code word, we choose one of these at random. If the corresponding $\mathbf{x}$ is a non-zero code word in D, then we have an undetectable error after decoding. Let $P_{\text{ue}}(D, C, p)$ be the probability that this happens. The question we must ask, how does use of the code C affect the probability of undetected error, that is, what is

$$R(D, C, p) \stackrel{\text{def}}{=} \frac{P_{\text{ue}}(D, C, p)}{P_{\text{ue}}(D, p)}.$$

By a careful choice of codes it is possible to get $R(D, C, p) \to 0$ when $p \to 0$. In particular, this is the case if $d_{\min}(C) > 2d_{\min}(D)$.

Remark 2.8. An alternative way of using the combined code is to correct up to t errors using code C and the use the remaining power to detect errors; this is a special case of the next method considered. In this case an error is undetected if the received vector is within distance t of a code word of the form $(\mathbf{x}|u(\mathbf{x}))$, where $\mathbf{x} \in D \setminus \{\mathbf{0}\}$, that is, within distance t of a non-zero code word in E. Hence the probability of undetected error is $P_{\text{ue}}^{(t)}(E, p)$.

$m \geq 1$

Let D be an $[mk, K; q]$ code and C an $[n, k; q]$ code which is able to correct t errors. Let E denote the concatenated code. Suppose that the all zero vector in $GF(q)^{mn}$ is transmitted. Let $(\mathbf{y}_1, \mathbf{y}_2, \cdots, \mathbf{y}_m)$ be received. The probability that $\mathbf{y}_l$ is within distance t of a fixed code word in C of weight i is given by

$$p'(i) = \sum_{j=0}^{n} N_t(i,j)\Big(\frac{p}{q-1}\Big)^j (1-p)^{n-j}.$$

If some $\mathbf{y}_l$ is at distance more than t from all code words in C, then a detectable error has occurred. Otherwise, each $\mathbf{y}_l$ is decoded to the closest code word $\tilde{\mathbf{y}}_l \in GF(q)^n$.

Let the corresponding information vector be $\tilde{\mathbf{z}}_l \in GF(q)^k$. If $(\tilde{\mathbf{z}}_1|\tilde{\mathbf{z}}_2|\cdots|\tilde{\mathbf{z}}_m)$ is a non-zero code word in D, then we have an undetectable error. The probability that this happens is then given by

$$\sum_{(i_1,i_2,\cdots,i_m)\neq(0,0,\cdots,0)} A_{i_1,i_2,\cdots,i_m}(C)p'(i_1)p'(i_2)\cdots p'(i_m).$$

2.10.3 *Probability of having received the correct code word after decoding*

The results in Section 2.9 can be generalized to situation where we combine correction of up to t errors and error detection. Assume that C is an $(n, M, d; q)$, that a code word from C is transmitted over a symmetric channel with symbol error probability p, and that the received vector is decoded to the closest code word if at most $t \leq (d-1)/2$ errors have occurred. The received vector is decoded into the sent code word if it is within Hamming distance of the sent code word. The probability for this to happen is $\sum_{j=0}^{t}\binom{n}{j}p^j(1-p)^{n-j}$. The probability that another code word is decoded to, by definition, $P_{\rm ue}^{(t)}(C,p)$. Hence the probability that we have decoded to the correct code word under the condition that we have been able to decode

$$P_{\rm corr}^{(t)}(C,p) = \frac{\sum_{j=0}^{t}\binom{n}{j}p^j(1-p)^{n-j}}{\sum_{j=0}^{t}\binom{n}{j}p^j(1-p)^{n-j} + P_{\rm ue}(t)(C,p)}.$$

It can be shown that $P_{\rm corr}^{(t)}(C,p)$ is strictly decreasing.

2.11 Complexity of computing $P_{\rm ue}(C,p)$

The following theorem shows that it is a hard problem to compute $P_{\rm ue}(C,p)$ in general. For a generator matrix G, let C_G denote the code generated by G.

Theorem 2.74. *The problem of computing $P_{\text{ue}}(C_G, p)$, as a function of a rational p and a generator matrix G, is an $\mathcal{NP}$ hard problem.*

Proof. It is known Berlekamp, McEliece, and van Tilborg (1978) that the following problem is $\mathcal{NP}$ complete:

> Given a $k \times n$ (binary) generator matrix G and an integer w, decide if the code C_G contains a code word of weight w.

In particular, this implies that the problem of finding the weight distribution of C_G is $\mathcal{NP}$ hard. We show that the problem of finding the weight distribution of C_G has the same complexity as the problem of evaluating $P_{\text{ue}}(C_G, p)$ in the sense that each has a polynomial time reduction to the other.

First, computing $P_{\text{ue}}(C_G, p)$, given a rational p and the weight distribution of C_G, is a simple evaluation of a polynomial using Theorem 2.1, and this can be done in polynomial time.

Next, if we know $P_{\text{ue}}(C_G, p_i)$ for n different values $p_1, p_2, \ldots, p_n$ (all different from 1), then the weight distribution can be determined in polynomial time from the set of n linear equation:

$$\begin{aligned}
z_1 A_1 + z_1^2 A_2 + \cdots + z_1^n A_n &= (1-p_1)^{-n} P_{\text{ue}}(C_G, p_1),\\
z_2 A_1 + z_q^2 A_2 + \cdots + z_q^n A_n &= (1-p_2)^{-n} P_{\text{ue}}(C_G, p_2),\\
\cdots\\
z_n A_1 + z_n^2 A_2 + \cdots + z_n^n A_n &= (1-p_n)^{-n} P_{\text{ue}}(C_G, p_n),
\end{aligned}$$

where $z_i = p_i/((q-1)(1-p_i))$. Since the coefficient matrix of this system of equations is a Vandermonde matrix, it has full rank and the set of equations determine $A_1, A_2, \ldots A_n$ uniquely. □

2.12 Particular codes

In this section we list the weight distributions of some known classes of codes. Results that apply only to binary codes are given in the next chapter in Section 3.5.

2.12.1 *Perfect codes*

Repetition codes and their duals

Over $GF(q)$ the (generalized) repetition codes are the $[n, 1; q]$ codes whose non-zero code words have Hamming weight n. The weight distribution function is

$$1 + (q-1)z^n.$$

Both the code and its dual are proper. In the binary case, the dual code is known as the single parity check code.

Hamming and simplex codes

For given m and q, the simplex code is the cyclic code over $GF(q)$ generated by a primitive polynomial of degree m. It has dimension m and length $n = \frac{q^m-1}{q-1}$. All non-zero code words have weight q^{m-1}, that is, its weight distribution function is

$$1 + \left(q^m - 1\right)z^{q^{m-1}}.$$

The dual code is the $\left[n, n-m; q\right]$ Hamming code. The weight distribution can be found from the weight distribution of the simplex code, using MacWilliams's identity:

$$A_i = \binom{\frac{q^m-1}{q-1}}{i}\frac{(q-1)^i}{q^m} + \frac{q^m-1}{q^m}\sum_{l=0}^{i}\binom{q^{m-1}}{l}\binom{\frac{q^{m-1}-1}{q-1}}{i-l}(-1)^l(q-1)^{i-l}.$$

Both codes are proper. The extended Hamming code and its dual code are also both proper.

Estimates of the minimum distances of shortened binary simplex codes (the partial periods of binary m-sequences) were given by Kumar and Wei (1992). Whether such shortened codes are good or not will depend on the primitive polynomial used for generating the code and the shortening done. Shortening of binary Hamming codes will be considered in more detail in the next chapter (in Section 3.4).

Golay codes

The binary Golay code is the $[23, 12, 7; 2]$ CRC code generated by the polynomial

$$z^{11} + z^{10} + z^6 + z^5 + z^4 + z^2 + 1.$$

Its weight distribution is given in Table 2.10. Both the code and its dual are proper.

The ternary Golay code is the $[11, 6, 5; 3]$ CRC code generated by the polynomial

$$z^5 + z^4 - z^3 + z^2 - 1.$$

Table 2.10 Weight distribution of the $[23, 12, 7; 2]$ Golay code.

i :	0	7	8	11	12	15	16	23
A_i :	1	253	506	1288	1288	506	253	1

Table 2.11 Weight distribution of the $[11, 6, 5; 3]$ Golay code.

i :	0	5	6	8	9	11
A_i :	1	132	132	330	110	24

Its weight distribution is given in Table 2.11. Both the code and its dual are proper.

References: MacWilliams and Sloane (1977, pp. 69, 482), Leung, Barnes, and Friedman (1979), Mallow, Pless, and Sloane (1976).

2.12.2 *MDS and related codes*

MDS codes

The Singleton bound says that $d \leq n - k + 1$ for an $[n, k, d; q]$ code. Maximum distance separable (MDS) codes are $[n, k, d; q]$ codes where $d = n - k + 1$. For these codes

$$A_i = \binom{n}{i} \sum_{j=0}^{i-n+k-1} (-1)^j \binom{i}{j} \left(q^{i-j-n+k} - 1\right)$$

for $n - k + 1 \leq i \leq n$, see Peterson and Weldon (1972, p. 72). Kasami and Lin (1984) proved that $P_{\rm ue}^{(t)}(C, p)$ is monotonous on $[0, \frac{q-1}{q}]$ for all $[n, k, d; q]$ MDS codes C and all $t < \frac{d}{2}$, that is, the codes are t-proper.

The defect and MMD codes

The *defect* $s(C)$ of an $[n, k, d; q]$ code is defined by

$$s = s(C) = n - k + 1 - d.$$

By the Singleton bound, $s \geq 0$, and C is MDS if and only if $s = 0$. Similarly, the *dual defect* $s^{\perp}(C)$ is the defect of the dual code, that is

$$s^{\perp} = s^{\perp}(C) = k + 1 - d^{\perp}.$$

Faldum and Willems (1997) proved that

$$A_{n-k+j} = \binom{n}{k-j} \sum_{i=0}^{j-s^\perp} (-1)^i \binom{n-k+j}{i} (q^{j-i}-1)$$
$$-(-1)^{j-s^\perp} \sum_{i=1}^{s+s^\perp-1} \binom{k+s-i}{k-j} \binom{j-1+s-i}{j-s^\perp} A_{n-k-s+i}.$$

Let C be an $[n,k,d;q]$ code with defect s. Olsson and Willems (1999) proved that if m is a positive integer and $k \geq m+1$, then

$$d \leq \frac{q^m(q-1)}{q^m-1}(s+m). \tag{2.49}$$

They called a code with equality in (2.49) a *maximum minimum distance code.* A complete characterization of MMD codes were done by Faldum and Willems (1998) and Olsson and Willems (1999). Based on this characterization, Dodunekova and Dodunekov (2002) showed that all MMD codes are proper, and Dodunekova (2003a) showed that the duals of all MMD codes are proper.

The MMD codes with $s \geq 1$ are the following codes (and any codes with the same weight distribution).

- S_k^{t*} for $t \geq 0$, where S_k is the simplex code.
- The $[q^{k-1}, k, (q-1)q^{k-2}; q]$ generalized Reed-Muller code of first order. For $q = 2$, we must have $k \geq 4$ (see e.g. Assmus and Key (1993)).
- The $[12, 6, 6; 3]$ extended Golay code.
- The $[11, 5, 6; 3]$ dual Golay code.
- The $[q^2, 4, q^2-q; q]$ projective elliptic quadratic code for $q > 2$ (see e.g. Dembowski, Finite Geometries, Springer, 1968).
- The $[(2^t-1)2^m + 2^t, 3, (2^t-1)2^m; 2^m]$ Denniston code for $1 \leq t \leq m$ (see Denniston (1969)).

Almost MDS codes

Codes with $s = 1$ are called *almost MDS* codes (AMDS). In particular, if also $d_2(C) = n-k+2$ (where $d_2(C)$ is the second minimum support weight defined in Section 1.5), the codes are called *near MDS* (NMDS). They have been discussed by de Boer (1996), Dodunekov and Landgev (1994), Faldum and Willems (1997). The error detecting properties of almost MDS codes were discussed Kløve (1995b) and Dodunekova, Dodunekov and Kløve

(1997). Some of these codes are good, others are not. For example, if C is an $[n, k, n-k; q]$ AMDS and

$$A_{n-k} \leq \frac{1}{q}\binom{n}{k}\left\{1 + 2\beta + 2\sqrt{\beta^2 + \beta}\right\},$$

where

$$\beta = \frac{1}{q} \cdot \frac{k-1}{k} \cdot \frac{n-k}{n-k+1},$$

then C is proper.

Expanded MDS codes

A symbol in $GF(q^m)$ can be represented as an m-tuple with elements in $GF(q)$. Therefore, an $[n, k; q^m]$ code can also be considered as an $[nm, km; q]$ code. For some Reed-Solomon codes (as well as some generalizations), the weight distribution of the corresponding binary codes have been studied by Imamura, Tokiwa, and Kasahara (1988), Kasami and Lin (1988), Kolev and Manev (1990), Pradhan and Gupta (1991), Retter (1991), and Retter (1992). In particular, Retter (1991) considered generalized Reed-Solomon codes generated by matrices of the form

$$\begin{pmatrix} v_1 & v_2 & \ldots & v_n \\ v_1\alpha_1 & v_2\alpha_2 & \ldots & v_n\alpha_n \\ v_1\alpha_1^2 & v_2\alpha_2^2 & \ldots & v_n\alpha_n^2 \\ \vdots & \vdots & \ddots & \vdots \\ v_1\alpha_1^{k-1} & v_2\alpha_2^{k-1} & \ldots & v_n\alpha_n^{k-1} \end{pmatrix}$$

where $\alpha_1, \alpha_2, \ldots, \alpha_n$ are distinct non-zero elements of $GF(q)$ and $v_1, v_2, \ldots, v_n$ are non-zero elements of $GF(q)$. Retter (1991) determined the average weight distribution for the class of codes where the α_i are kept fixed, but the v_i is varied in all possible $(q-1)^n$ ways. Nishijima (2002) considered upper bounds on the average probability of undetected error for this class of codes, and Nishijima (2006) gave upper and lower bounds on the probability of undetected error for individual codes in the class.

2.12.3 *Cyclic codes*

Several of the codes presented under other headings are cyclic. Here we will give some references to papers dealing with cyclic codes in general, and more particular, irreducible cyclic codes.

Barg and Dumer (1992) gave two algorithms for computing the weight distribution of cyclic codes.

A cyclic code is called irreducible if the polynomial $h(z)$ in Theorem 1.2 is irreducible. An irreducible cyclic $[n, k; q]$ code C can be represented as follows.

$$\Big\{\Big(\mathrm{Tr}_1^k(x), \mathrm{Tr}_1^k(x\beta), \ldots, \mathrm{Tr}_1^k(x\beta^{n-1})\Big) \;\Big|\; x \in GF(q^k)\Big\},$$

where Tr is the trace function and β is a root of $h(z) = 0$.

McEliece and Rumsey (1972) and Helleseth, Kløve, and Mykkeltveit (1977) gave general methods for computing the weight distribution of irreducible cyclic codes and also gave explicit expressions for several infinite classes of such codes. Their papers also contain a number of further references. Segal and Ward (1986) and Ward (1993) have computed the weight distribution of a number of irreducible cyclic codes.

One drawback with the use of cyclic codes is that they are not good for block synchronization. However, we can overcome this drawback by using some proper coset S of the cyclic code C since $P_{\mathrm{ue}}(C, p) = P_{\mathrm{ue}}(S, p)$.

2.12.4 *Two weight irreducible cyclic codes*

Baumert and McEliece (1972) and Wolfmann (1975) studied a class of q-ary two weight linear irreducible cyclic codes. The class is parametrized by three parameters, $r \geq 1$, $t \geq 2$, and $s \geq 2$, where s divides $q^r + 1$. The dimension and length of the code are

$$k = 2rt, \quad n = \frac{q^k - 1}{s}.$$

The codes have the two weights and corresponding number of code words:

$$\begin{aligned} w_1 &= (q-1)\tfrac{q^{k-1}+(-1)^t(s-1)q^{rt-1}}{s}, & A_{w_1} &= n, \\ w_2 &= (q-1)\tfrac{q^{k-1}-(-1)^t q^{rt-1}}{s}, & A_{w_2} &= n(s-1). \end{aligned}$$

The error detecting properties for these codes were studied by Dodunekova, Rabaste, Paez (2004) and Dodunekova and Dodunekov (2005a). They showed that the codes and their duals are proper in the following cases:

$$\begin{aligned} &q = 2,\ s \geq 2,\ t \text{ even},\ r \geq 1, \\ &q = 2,\ s = 3,\ t \text{ odd},\ r \geq 1, \\ &q = 3,\ s = 2,\ t \geq 2,\ r \geq 1, \\ &q = 3,\ s = 4,\ t = 2,\ r = 1. \end{aligned}$$

In all other cases, both the codes and their duals are bad.

2.12.5 *The product of two single parity check codes*

Leung (1983) determined $A_C(z)$ for the product of two binary single parity check codes. If C_1 and C_2 are $[l, l-1]$ and $[m, m-1]$ parity check codes respectively, then the weight distribution function of the product code is

$$\frac{1}{q^{l+m-1}} \sum_{i=0}^{m-1} \binom{m-1}{i} \left(z^i + z^{m-i}\right)^l.$$

The product code is proper for

$$(l, m) \in \{(2,2), (2,3), (2,4), (2,5), (3,2), (3,3), (3,4), (3,5), (4,2), (4,3), (4,4), (5,2), (5,3)\},$$

in all other cases the product code is bad.

2.13 How to find the code you need

There are two main classes of problems we may encounter when we want to use an error detecting code on the q-SC:

- how to calculate $P_{\text{ue}}(C,p)$ or $P_{\text{ue}}^{(t)}(C,p)$ for a given code,
- how to find an $(n, M; q)$ code or $[n, k; q]$ code for which $P_{\text{ue}}(C,p)$ or $P_{\text{ue}}^{(t)}(C,p)$ is below some given value.

In general these are difficult problems and a solution may be infeasible. However, in many cases of practical interest, solutions may be found using the results and methods given in this and the next chapter. Since the results are many and diverse, we give in this section a short "handbook" on how to attack these problems.

Given C and p, how to find or estimate $P_{\text{ue}}(C,p)$

The first thing to do is to check in Sections 2.12 and 3.5 if your code is in the lists of codes with known weight distribution. If not, the weight distribution of the code or its dual may possibly be found by a complete listing of the code words (if k or $n-k$ are small). Having found the weight distribution, $P_{\text{ue}}(C,p)$ can be computed from Theorem 2.1. If you have found the weight distribution of the dual code, $P_{\text{ue}}(C,p)$ can be found combining Theorems 2.1 and 2.7. Likewise, $P_{\text{ue}}^{(t)}(C,p)$ can be computed from (2.47) if the weight distribution is known, and from Theorem 1.14 and (2.47) if the weight distribution of the dual code is known.

If the weight distribution is not possible to obtain, you have to be satisfied with estimates for $P_{\text{ue}}(C,p)$. One upper bound is given in Theorem 2.51. Another is obtained by combining Theorems 1.22, 1.23 and 2.50. Any partial information you may have about the weight distribution may help to improve the bound thus obtained. If $p \leq d/n$, Theorem 2.2 can be used. Lower bounds may be found using Theorems 2.42–2.48. Upper bounds on $P_{\text{ue}}^{(t)}(C,p)$ is obtained by combining (2.47) and Theorems 1.22 and 1.23.

Given C, a, and b, how to find or estimate $P_{\text{wc}}(C,a,b)$

To determine $P_{\text{wc}}(C,a,b)$ exactly is usually a harder problem than to determine $P_{\text{ue}}(C,p)$ for a particular p. In the special cases when we can prove that the code is proper, and $b \leq (q-1)/q$, then $P_{\text{wc}}(C,a,b) = P_{\text{ue}}(C,b)$. Likewise, if $b \leq d/n$, then $P_{\text{wc}}(C,a,b) = P_{\text{ue}}(C,b)$ by Theorem 2.2. If the code is good, then by definition $P_{\text{wc}}(C,0,(q-1)/q) = P_{\text{ue}}(C,(q-1)/q) = (M-1)/q^n$.

With known weight distribution of the code, the algorithm based on Theorem 2.35 can be used to get as good an approximation to $P_{\text{wc}}(C,a,b)$ as we want. Otherwise, any upper bounds on the weight distribution (e.g. from Theorems 1.22 and 1.23) can be combined with Theorem 2.35 to give upper bounds on $P_{\text{wc}}(C,a,b)$. The last resort is to hope that $P_{\text{wc}}(C,a,b)$ is about average and that the upper bound on the average (2.41) is also an upper bound on $P_{\text{wc}}(C,a,b)$ (this is not very satisfactory, cf. Theorem 2.39!).

Given p and a bound B, how to find an $(n,M;q)$ code C such that $P_{\text{ue}}(C,p) \leq B$

It is possible that B is too small, that is, there are no $(n,M;q)$ codes C with the required property. The first thing to do is therefore to compare B with the lower bounds in Theorems 2.42–2.48. If B is not smaller than any these lower bounds, the next thing to do is to look through the list of codes in Sections 2.12 and 3.5 to see if any of those satisfy the requirement. If not, possibly a code with the requirements can be obtained using the *-operation one or more times on any of these codes. If k or $n-k$ is not too large, the next possibility is to pick a number of $(n,M;q)$ or $[n,k;q]$ codes at random and check them.

The problem of finding the $[n,k;q]$ code which minimizes $P_{\text{ue}}(C,p)$ is usually harder, but the line of attack would be the same. The solution for $k \leq 4$ and for $n-k \leq 4$ is given in Section 3.2. Further, if $n =$

$s(q^k-1)/(q-1)$, then the optimal code for all p is $C^{(s-1)*}$, where C is the $[(q^k-1)/(q-1), k; q]$ simplex code. Further, the $[(q^m-1)/(q-1), (q^m-1)/(q-1)-m; q]$ Hamming codes are also optimal for all p.

Given a, b and a bound B, how to find an $(n, M; q)$ code C such that $P_{\mathrm{we}}(C, a, b) \le B$

This problem should be attacked in the same way as the previous problem, the main difference is that now it is $P_{\mathrm{we}}(C, a, b)$ which needs to be computed or estimated. As for a lower bound, note that $P_{\mathrm{we}}(C, a, b) \ge (M-1)/q^n$ if $a \le (q-1)/q \le b$.

How to choose a code which can be used for different sizes of the information, but with a fixed number of check bits

The problem can be stated as follows: find an $[m+r, r; q]$ code such that the shortened $[k+r, k; q]$ code is good for all k within some range, possibly for $1 \le k \le m$. The main choices of codes for this problem are CRC codes. You should consult the sections on CRC codes and, if need be, the references given there.

2.14 The local symmetric channel

A channel is called a local symmetric channel (LSC) if it behaves as a q-ary symmetric channel for each transmitted symbol, but the channel error probability may vary from symbol to symbol. We consider linear codes over $GF(q)$. Suppose that an $[n, k; q]$ code C is used on a LSC where the probability that symbol i is in error is given by p_i. A similar argument as for the q-ary symmetric channel gives the following expression for the probability of undetected error.

$$
\begin{aligned}
&P_{\mathrm{ue}}(C, p_1, p_2, \cdots, p_n) \\
&= \prod_{i=1}^{n}(1-p_i)\Big\{A_C\Big(\frac{p_1}{(q-1)(1-p_1)}, \cdots, \frac{p_n}{(q-1)(1-p_n)}\Big) - 1\Big\} \\
&= \frac{M}{q^n} A_{C^\perp}\Big(1-\frac{q}{q-1}p_1, \cdots, 1-\frac{q}{q-1}p_n\Big) - \prod_{i=1}^{n}(1-p_i).
\end{aligned}
$$

Example. Let C be the $[n, n-1; 2]$ even-weight code, that is, $C^\perp = \{\mathbf{0}, \mathbf{1}\}$.

Then

$$A_{C^\perp}(z_1, z_2, \cdots, z_n) = 1 + z_1 z_2 \cdots z_n,$$

and so

$$P_{\rm ue}(C, p_1, p_2, \cdots, p_n) = \frac{1}{2}\Big\{1 + \prod_{i=1}^{n}(1 - 2p_i)\Big\} - \prod_{i=1}^{n}(1 - p_i).$$

In analogy to what we did for the q-SC, we define

$$P_{\rm wc}(C, a, b) = \max\Big\{P_{\rm ue}(p_1, p_2, \ldots, p_n) \,\Big|\, p_i \in [a, b] \text{ for } 1 \le i \le n\Big\}.$$

Theorem 2.75. *Let C be an $[n, k; q]$ code, and let $0 \le a \le b \le 1$. For $X \subseteq \{1, 2, \ldots, n\}$, define $\mathbf{z}(X) = \mathbf{z}$ by*

$$z_i = \begin{cases} b \text{ if } i \in X, \\ a \text{ if } i \notin X. \end{cases}$$

Then

$$P_{\rm wc}(C, a, b) = \max\Big\{P_{\rm ue}(C, \mathbf{z}(X)) \,\Big|\, X \subseteq \{1, 2, \ldots, n\}\Big\}.$$

Proof. For convenience, we write

$$Q = \{\mathbf{p} \mid a \le p_i \le b \text{ for } 1 \le i \le n\}.$$

We prove by induction on j that there exists a vector $\mathbf{y} \in Q$ such that

$$y_i \in \{a, b\} \quad \text{for} \quad 1 \le i \le j, \tag{2.50}$$

$$P_{\rm ue}(C, \mathbf{p}) \le P_{\rm ue}(C, \mathbf{y}) \text{ for all } \mathbf{p} \in Q. \tag{2.51}$$

First, let $j = 0$. Since Q is a closed set and $P_{\rm ue}(C, \mathbf{p})$ is a continuous function of $\mathbf{p}$ on Q, it obtains its maximum on Q, that is, there exists a $\mathbf{y} \in Q$ such that (2.51). Moreover, (2.50) is trivially true for $j = 0$. This proves the induction basis.

For the induction step, let $j > 0$ and suppose that the statement is true for $j - 1$; let $\mathbf{y} \in Q$ such that $y_i \in \{a, b\}$ for $1 \le i \le j - 1$ and (2.51) is satisfied. For a $\mathbf{c} \in C$ we have

$$\prod_{i=1}^{n}\Big(\frac{y_i}{q-1}\Big)^{w_{\rm H}(c_i)}(1 - y_i)^{1 - w_{\rm H}(c_i)}$$

$$= \begin{cases} \frac{y_j}{q-1}\prod_{i \ne j}\Big(\frac{y_i}{q-1}\Big)^{w_{\rm H}(c_i)}(1 - y_i)^{1 - w_{\rm H}(c_i)} & \text{if } c_j \ne 0, \\ (1 - y_j)\prod_{i \ne j}\Big(\frac{y_i}{q-1}\Big)^{w_{\rm H}(c_i)}(1 - y_i)^{1 - w_{\rm H}(c_i)} & \text{if } c_j = 0. \end{cases}$$

Hence

$$P_{\text{ue}}(C, \mathbf{y}) = \alpha_j y_j + \beta_j$$

where

$$\alpha_j = \frac{1}{q-1} \sum_{\substack{\mathbf{c}\in C \\ c_j \neq 0}} \prod_{i \neq j} \Big(\frac{y_i}{q-1}\Big)^{w_{\text{H}}(c_i)} (1-y_i)^{1-w_{\text{H}}(c_i)}$$
$$- \sum_{\substack{\mathbf{c}\in C\setminus\{\mathbf{0}\} \\ c_j = 0}} \prod_{i \neq j} \Big(\frac{y_i}{q-1}\Big)^{w_{\text{H}}(c_i)} (1-y_i)^{1-w_{\text{H}}(c_i)}$$

$$\beta_j = \sum_{\substack{\mathbf{c}\in C\setminus\{\mathbf{0}\} \\ c_j = 0}} \prod_{i \neq j} \Big(\frac{y_i}{q-1}\Big)^{w_{\text{H}}(c_i)} (1-y_i)^{1-w_{\text{H}}(c_i)}.$$

If $\alpha_j > 0$ we must have $y_j = b$ since $\mathbf{y}$ is a maximum point, and so (2.50) is true. Similarly, if $\alpha_j < 0$ we must have $y_j = a$ and again (2.50) is true. Finally, if $\alpha_j = 0$, then $P_{\text{ue}}(C, \mathbf{y})$ is independent of the value of y_j. Therefore, we may replace y_j by any other value in $[a, b]$, e.g. a which makes (2.50) true also in this case. This proves the statement for j and the induction is complete. For $j = n$ we see that (2.50) and (2.51) implies that there exists an X such that $P_{\text{wc}}(C, a, b) = P_{\text{ue}}\Big(C, \mathbf{z}(X)\Big)$. □

Lemma 2.9. *If $a = 0$ and $b = 1$, then*

$$P_{\text{ue}}\Big(C, \mathbf{z}(X)\Big) = \frac{\#\{\mathbf{y} \in C \mid \chi(\mathbf{y}) = X\}}{(q-1)^{\#X}}.$$

Proof. In the case we consider, $\mathbf{z}(X)$ is given by

$$z_i = \begin{cases} 1 \text{ if } i \in X, \\ 0 \text{ if } i \notin X \end{cases}$$

and so

$$\Big(\frac{z_i}{q-1}\Big)^{w_{\text{H}}(c_i)} (1-z_i)^{1-w_{\text{H}}(c_i)} = \begin{cases} 1 & \text{if } w_{\text{H}}(c_i) = z_i = 0 \text{ and } c_i = c_i', \\ 1/(q-1) & \text{if } w_{\text{H}}(c_i) = z_i \neq 0 \text{ and } c_i \neq c_i', \\ 0 & \text{otherwise.} \end{cases}$$

Hence

$$\prod_{i=1}^{n} \Big(\frac{z_i}{q-1}\Big)^{w_{\text{H}}(c_i)} (1-z_i)^{1-w_{\text{H}}(c_i)} = \begin{cases} 1/(q-1)^{\#X} & \text{if } \chi(\mathbf{c}) = X, \\ 0 & \text{otherwise.} \end{cases}$$

Summing over all $\mathbf{c} \in C \setminus \{\mathbf{0}\}$, the lemma follows. □

Theorem 2.76. *Let C be an $[n,k,d;q]$ code. Then*

$$P_{\rm wc}(C,0,1)=\frac{1}{(q-1)^{d-1}}.$$

Proof. First, let $\mathbf{c}$ be a code word in C of minimum weight d, and let $X=\chi(\mathbf{c})$. Then $\chi(a\mathbf{c})=X$ for all $a\neq 0$. Hence

$$P_{\rm wc}(C,0,1)\geq P_{\rm ue}\Big(C,\mathbf{z}(X)\Big)\geq\frac{q-1}{(q-1)^d}=\frac{1}{(q-1)^{d-1}}. \tag{2.52}$$

On the other hand, let X be the support of some non-zero code word. Let

$$B_X=\{\mathbf{c}_X\mid \mathbf{c}\in C \text{ and } \chi(\mathbf{c})=X\}$$

where $\mathbf{c}_X$ denotes the vector of length $\#X$ obtained by puncturing all positions of $\mathbf{c}$ not in X (the elements in the punctured positions are all zero). Let U_X be the vector space generated by B_X. Let r be the dimension and δ the minimum distance of U_X. Clearly, $d\leq\delta$. By the Singleton bound we have $\delta\leq\#X-r+1$. Hence $r\leq\#X-(d-1)$. By Theorem 1.23, U_X contains not more than $(q-1)^{\#X-(d-1)}$ code words of weight $\#X$. In particular, $\#B_X\leq(q-1)^{\#X-(d-1)}$. By Lemma 2.9 we have

$$P_{\rm ue}\Big(C,\mathbf{z}(X)\Big)=\frac{\#B_X}{(q-1)^{\#X}}\leq\frac{(q-1)^{\#X-(d-1)}}{(q-1)^{\#X}}=\frac{1}{(q-1)^{d-1}}.$$

By Theorem 2.75 we have

$$P_{\rm wc}(C,0,1)=\max\Big\{P_{\rm ue}\Big(C,\mathbf{z}(X)\Big)\;\Big|\;X\subseteq\{1,2,\ldots,n\}\Big\}\leq\frac{1}{(q-1)^{d-1}}.$$

This, together with (2.52), proves the theorem. □

Lemma 2.10. *Consider the interval $[0,(q-1)/q]$. We have*

$$P_{\rm ue}\Big(C,\mathbf{z}(X)\Big)=\frac{\#\{\mathbf{y}\in C\setminus\{\mathbf{0}\}\mid\chi(\mathbf{y})\subseteq X\}}{q^{\#X}}.$$

Proof. First we note that if $i\in X$, then $z_i=(q-1)/q$, and so

$$\frac{z_i}{q-1}=1-z_i=\frac{1}{q}.$$

Therefore

$$\Big(\frac{z_i}{q-1}\Big)^{w_{\rm H}(c_i)}(1-z_i)^{1-w_{\rm H}(c_i)}=\frac{1}{q}$$

for any value of c_i. If $i\notin X$, then $z_i=0$ and so

$$\Big(\frac{z_i}{q-1}\Big)^{w_{\rm H}(c_i)}(1-z_i)^{1-w_{\rm H}(c_i)}=\begin{cases}1 \text{ if } c_i=0,\\ 0 \text{ if } c_i\neq 0.\end{cases}$$

Hence

$$\prod_{i=1}^{n}\Big(\frac{z_i}{q-1}\Big)^{w_{\rm H}(c_i)}(1-z_i)^{1-w_{\rm H}(c_i)}=\begin{cases}q^{-\#X} & \text{if } \chi(\mathbf{y})\subseteq X,\\ 0 & \text{if } \chi(\mathbf{y})\not\subseteq X.\end{cases}$$

Summing over all non-zero $\mathbf{y}$ in C, the lemma follows. □

For $X \subseteq \{1, 2, \ldots, n\}$, let

$$V_X = \{\mathbf{c} \in C \mid \chi(\mathbf{c}) \subseteq X\}.$$

We note that V_X is a vector space, i.e. V_X is a subcode of C. If the dimension of V_X is r, then

$$\#\Big\{\mathbf{y} \in C \setminus \{\mathbf{0}\} \mid \chi(\mathbf{y}) \subseteq X\Big\} = q^r - 1$$

and

$$d_r \leq \#\chi(V_X) \leq \#X.$$

Hence

$$P_{\text{ue}}\Big(C, \mathbf{z}(X)\Big) \leq \frac{q^r - 1}{q^{d_r}}. \tag{2.53}$$

Moreover, for each r, $1 \leq r \leq k$, there exists a set X such that $\#X = d_r$ and $\dim(V_X) = r$. Combining this with Theorem 2.75 and Lemma 2.10 we get the following result.

Theorem 2.77. *For an $[n, k; q]$ code C we have*

$$P_{\text{wc}}(C, 0, (q-1)/q) = \max_{1 \leq r \leq k} \left\{\frac{q^r - 1}{q^{d_r}}\right\}.$$

Next, we state a lemma whose simple proof is omitted.

Lemma 2.11. *Let l, m, and t be positive integers. Then*

$$(i) \quad \frac{q^l - 1}{q^m} < \frac{q^{l+t} - 1}{q^{m+t}}, \tag{2.54}$$

$$(ii) \quad \frac{q^l - 1}{q^m} > \frac{q^{l+t} - 1}{q^{m+u}} \quad \textit{for } u \geq t + 1. \tag{2.55}$$

We can now restate Theorem 2.77.

Theorem 2.78. *Let C be an $[n, k, d; q]$ code. Let*

$$s = \max\Big\{r \;\Big|\; 1 \leq r \leq k \text{ and } d_r = d_1 + (r-1)\Big\}.$$

Then

$$P_{\text{wc}}(C, 0, (q-1)/q) = \frac{q^s - 1}{q^{d+s-1}}.$$

Proof. First, since $d_s - d_r = s - r$ for $1 \le r \le s$, we have, by Lemma 2.11 (i),

$$\frac{q^r - 1}{q^{d_r}} < \frac{q^s - 1}{q^{d_s}}$$

for $1 \le r < s$. For $s < r \le k$ we have, by (1.18), that $d_r - d_s \ge r - s + 1$, and by Lemma 2.11 (ii) we get

$$\frac{q^s - 1}{q^{d_s}} > \frac{q^r - 1}{q^{d_r}}.$$

Hence, by Theorem 2.77, we get

$$P_{\text{wc}}\Big(C, \frac{q-1}{q}\Big) = \frac{q^s - 1}{q^{d_s}} = \frac{q^s - 1}{q^{d+s-1}}.$$

□

We give some corollaries of Theorem 2.78. Usually only part of the weight hierarchy is needed to determine $P_{\text{wc}}(C, 0, (q-1)/q)$, sometimes only d_1. For instance, by (1.17) we have $d_2 > d_1 + 1$ if $d_1 \ge q + 1$. Hence we have the following corollary.

Corollary 2.15. *Let C be an $[n, k, d; q]$ code with minimum distance $d > q$. Then*

$$P_{\text{wc}}\Big(C, 0, \frac{q-1}{q}\Big) = \frac{q-1}{q^d}.$$

From Theorem 2.77 we see that

$$P_{\text{wc}}\Big(C, 0, \frac{q-1}{q}\Big) \ge \frac{q^k - 1}{q^{d_k}} \ge \frac{q^k - 1}{q^n}.$$

In particular, $P_{\text{wc}}(C, (q-1)/q) = (q^k - 1)/q^n$ if and only if $n = d_k = d_1 + k - 1$, that is, if and only if the code is MDS.

Corollary 2.16. *For an $[n, k; q]$ code C we have*

$$P_{\text{wc}}\Big(C, 0, \frac{q-1}{q}\Big) \ge \frac{q^k - 1}{q^n}$$

with equality if and only if C is an MDS code.

Remark 2.9. The average value of $P_{\text{ue}}(C, p_1, p_2, \ldots, p_n)$ over all $[n, k; q]$ codes can be shown to be

$$E\Big(P_{\text{ue}}(C, p_1, p_2, \ldots, p_n)\Big) = \frac{q^k - 1}{q^n - 1}\Big\{1 - \prod_{i=1}^{n}(1 - p_i)\Big\},$$

and so

$$\max\Big\{E\Big(P_{\text{ue}}(C, p_1, p_2, \ldots, p_n)\Big) \;\Big|\; 0 \le p_i \le (q-1)/q\Big\} = \frac{q^k - 1}{q^n}.$$

In analogy to what we did for the q-SC, we could call a code good if $P_{\rm wc}(C,0,(q-1)/q) = (q^k-1)/q^n$. However, we note that this is a very strong condition, which by Corollary 2.16 is satisfied only by MDS codes.

It is often difficult to determine the weight hierarchy. Therefore it is useful to have good bounds. Corollary 2.16 gives such a lower bound. Theorem 2.78 and Lemma 2.11 a) give the following general bounds.

Corollary 2.17. *Let C be an $[n,k,d;q]$ code. Then*

$$\frac{q-1}{q^d} \le P_{\rm wc}\Big(C,0,\frac{q-1}{q}\Big) \le \frac{q^k-1}{q^{d+k-1}} = \frac{q-q^{-(k-1)}}{q^d}.$$

2.15 Comments and references

2.1. Most of the material in this section is basic and it has been introduced by various people over the years in various notations.
Example 2.2 is due to Honkala and Laihonen (1999).
Theorem 2.2 is an observation done by several people.
The threshold was introduced by Kløve (1996b).

2.2. The main theorem 2.7 follows directly from the results given in Chapter 1. It was first given by Leontev (1972) (for $q=2$).
Theorem 2.8 in the special case of binary linear codes was presented in Kløve (1987) and published in Kløve and Korzhik (1995).

2.3. Sturm sequences are treated in many introductory texts on numerical mathematics, e.g. Isaacson and Keller (1966).
Theorem 2.9 is a special case of a method given in Kløve (1984c). For the binary case, it was given in Kløve and Korzhik (1995).
Theorem 2.10 is essentially due to Perry (1991). The special case with $i=d$ is due to Korzhik and Fink (1975, page 184) (it was rediscovered by Padovani and Wolf (1982)). A similar theorem for 1-good codes was given by Lin (1990).
Theorem 2.11 is also due to Perry (1991).
Theorem 2.12 was presented in Kløve (1987).
Theorem 2.13 and the corollaries are based on Perry and Fossorier (2006a) and Perry and Fossorier (2006b).
The results on $\mu(d,k)$, in particular Theorem 2.14, were given by Naydenova and T. Kløve (2005a) and Naydenova and T. Kløve (2005b).
Theorem 2.16 was given in Nikolova (2005a) and Dodunekova and Nikolova (2005a).

A special case of Theorem 2.17 was given by Dodunekova and Nikolova (2005a). They also proved the claim in Example 2.9 in the binary case. Theorems 2.18 and 2.19 and Lemma 2.5 are due to Dodunekova and Dodunekov (1997b).

Theorem 2.20 is due to Naydenova and T. Kløve (2006a).

Theorem 2.21 in the binary case was essentially shown by Fu and Kløve (2006) (the report considered *good* binary linear codes, and codes that are not good are also not proper).

2.4. Theorem 2.23 for the binary case was given in Kløve and Korzhik (1995). The generalization to non-linear codes, Theorem 2.22, is new here. However, the basic idea is old.

Theorem 2.25 for the binary case was given in Kløve and Korzhik (1995). The generalization to non-linear codes, Theorem 2.24, is new here.

Theorem 2.28 is due to Leontev (1972). Corollary 2.8 is due to Levenshtein (1978). The generalizations of both to non-linear codes, (Theorem 2.26 and Corollary 2.7) are new here.

Theorem 2.27 is due to Nikolova (2005b).

Theorems 2.30 and 2.31 are due to Korzhik (1965) (for $q = 2$). An upper bound on $\alpha(\mathbf{x})$ due to Lin (1990) is identical to (2.27) for $d = 3$, weaker for $d > 3$.

Theorems 2.32, 2.33, and 2.34 were given in Kløve and Korzhik (1995) for the binary case.

2.5. The bounds on $P_{\mathrm{wc}}(\mathcal{C}, a, b)$ (for binary linear codes) are from Kløve (1994a) and Kløve (1995a). Corollary 2.11 is due to Massey (1978). He also gave the bound $S_n \leq n$.

The sums $S_{n,k}$ were studied by Szpankowski (1995). In particular, (2.38) and (2.39) are due to him. The bounds (2.37) are from Kløve (1994a).

Theorems 2.37 and 2.38 were given by Blinovsky (1995), Blinovsky (1996).

Theorem 2.39 is a special case of a theorem given in Kløve (1984a), the example to prove it given here is simpler. However, Kløve (1984a) gives the stronger result that there exist codes C with arbitrary large minimum distance and the property given in Theorem 2.39.

2.6. Theorem 2.42 is essentially due to Korzhik (1965).

Theorem 2.43 is due to Leontev (1972).

Theorem 2.45 is due to AbdelGhaffar (1997). A simpler proof of the slightly weaker bound in Theorem 2.44 was given by Fu, Kløve, and Wei (2003).

Theorem 2.46 is due to Kløve (1996c). A similar, but weaker bound was given by Wolf, Michelson, and Levesque (1982).
Theorems 2.50 and 2.51 were given in Kløve and Korzhik (1995) (in the binary case). A weaker upper bound based on the same idea was given in Kasami, Kløve, and Lin (1983).
Theorem 2.52 is due to Leontev (1972).
Theorems 2.53 and 2.54 are due to Levenshtein (1978). A more general bound, which has the bounds in Theorems 2.52 and 2.53 as special cases, was given by Katsman (1994).
Theorems 2.48 and 2.55 are due to Levenshtein (1989).
Theorems 2.57 and 2.58 are due to Levenshtein (1978) and Levenshtein (1989).
Theorem 2.56 is due to Ashikhmin and Barg (1999).
Theorem 2.59 is due to Barg and Ashikhmin (1999)

2.7. Theorems 2.61 and 2.62 are improved versions, with new proofs of results given in Kløve and Korzhik (1995).

2.8. The results and examples presented are mainly from Kløve and Korzhik (1995). However, other examples to illustrate some of these results have been given by Leung (1983), Leung and Hellman (1976), Leung, Barnes, and Friedman (1979).

2.9. The function $P_{\text{corr}}(C,p)$ was studied Faldum (2005) and Faldum, Lafuente, Ochoa, and Willems (2006). They used the notation $P_{\text{fd}}(C,0,p) = 1 - P_{\text{corr}}(C,p)$.
Theorem 2.67 is due to Faldum (2005).
Theorem 2.70 is due to Ashikhmin (private communication 2006).

2.10. A similar discussion of $P_{\text{ue}}^{(t)}(C,p)/P_{\text{ue}}(C,p)$ for the binary case was done by Kløve (1984a).
For more details on the concatenated codes with $m = 1$ we refer to Kløve and Miller (1984).
Bounds on the probability of undetected error for a concatenated code with Reed-Solomon outer code are given in Korzhik and Fink (1975, pages 188-190).
For the concatenated codes with $m > 1$, see Kasami, Fujiwara, and Lin (1986) which have a further discussion and some examples.
Many systems use a concatenation of a convolutional code for error correction and a block code (usually a CRC code) for error detection. Some papers discussing various such systems are Harvey and Wicker (1994), Seshadri and Sundberg (1994), Yamamoto and Itoh (1980).

2.11. The observation that the problem of computing $P_{\text{ue}}(C,p)$ in general is an $\mathcal{NP}$ hard problem was given in Kløve and Korzhik (1995).

2.14. The results on the worst-case probability of undetected error on the LSC are generalizations to general q of results given in the binary case by Kløve (1996a).

Chapter 3

Error detecting codes for the binary symmetric channel

For $q = 2$, the q-ary symmetric channel is called the binary symmetric channel (BSC). In this chapter we give some results that apply for the BSC, but which does not generalize to general q or a possible generalization is not known or has not been studied. Since $q = 2$ throughout the chapter, we (usually) drop ";2" from the notations and write $[n,k]$ for $[n,k;2]$, etc.

3.1 A condition that implies "good"

For linear codes with code words of even weight only, the following theorem is sometimes useful to show that

$$P_{\text{ue}}(C,p) \leq 2^{k-n}\{1 + (1-2p)^n - 2(1-p)^n\}. \tag{3.1}$$

Since $1 + (1-2p)^n - 2(1-p)^n$ is monotonically increasing on $[0, 1/2]$, this is a stronger condition than C being good.

Since all code words have even weight, the all-one vector $\mathbf{1}$ belongs to $C^\perp$. Therefore, for any code word $\mathbf{x} \in C^\perp$ of weight i, there is a code word in $C^\perp$ of weight $n-i$, namely $\mathbf{x}+\mathbf{1}$. Hence we can write the weight distribution function of $C^\perp$ as follows:

$$A_{C^\perp}(z) = \sum_{i=0}^{\nu} B_i(z^i + z^{n-i}),$$

where $\nu = \lfloor \frac{n}{2} \rfloor$.

Remark 3.1. If n is even, then $B_\nu = \frac{1}{2}A_\nu(C^\perp)$. For $i < \frac{n}{2}$ we have $B_i = A_i(C^\perp)$.

Theorem 3.1. *Let C be an $[n,k,d]$ code where $n = 2\nu + 1$ is odd and all code words have even weight. If*

$$(2^{n-k} - 2)\binom{\nu}{j} \geq 2^{2j+1} \sum_{i=j}^{\nu-1} \binom{i+j}{2j} B_{\nu-i}$$

for

$$\frac{d}{2} \leq j < \frac{(n - 2d^{\perp})^2 + n}{2n},$$

then

$$P_{\text{ue}}(C,p) \leq 2^{k-n} \{1 + (1-2p)^n - 2(1-p)^n\}$$

for all $p \in \left[0, \frac{1}{2}\right]$.

Theorem 3.2. *Let C be an $[n,k,d]$ code where $n = 2\nu$ is even and all code words have even weight. If*

$$(2^{n-k} - 2)\binom{\nu}{j} \geq 2^{2j} \sum_{i=j}^{\nu-1} \left\{ \binom{i+j}{2j} + \binom{i+j-1}{2j} \right\} B_{\nu-i}$$

for

$$\frac{d}{2} \leq j < \frac{(n - 2d^{\perp})^2 + n - 2}{2n - 2},$$

then

$$P_{\text{ue}}(C,p) \leq 2^{k-n} \{1 + (1-2p)^n - 2(1-p)^n\}$$

for all $p \in \left[0, \frac{1}{2}\right]$.

Proof. We sketch the proof of Theorem 3.1. By Theorem 2.4

$$\begin{aligned} P_{\text{ue}}(C,p) &= 2^{n-k} A_{C^{\perp}}(1-2p) - (1-p)^n \\ &= 2^{k-n} \{1 + (1-2p)^n - 2(1-p)^n - F(p)\} \end{aligned}$$

where

$$F(p) = (2^{n-k} - 2)(1-p)^n - \sum_{i=1}^{\nu} B_i \left\{(1-2p)^i + (1-2p)^{n-i}\right\}.$$

We will show that $F(p) \geq 0$ for all p. We see that

$$(1-p)^n = (1-p) \sum_{j=0}^{\nu} \binom{\nu}{i} p^{2j} (1-2p)^{\nu-j}$$

and

$$(1-2p)^i + (1-2p)^{n-i} = (1-p)\sum_{j=0}^{i} 2^{2j+1}\binom{i+j}{2j} p^{2j}(1-2p)^{\nu-j}.$$

Hence

$$F(p) = (1-p)\sum_{j=0}^{\nu} p^{2j}(1-2p)^{\nu-j}F_j,$$

where

$$F_j = (2^{n-k}-2)\binom{\nu}{j} - \sum_{i=j}^{\nu-1} 2^{2j+1}\binom{i+j}{2j} B_{\nu-i}.$$

We will show that $F_j \geq 0$ for all j. By assumption, this is true for $\frac{d}{2} \leq j < d_0 \stackrel{\text{def}}{=} \left\lceil \frac{(n-2d^\perp)^2+n}{2n} \right\rceil$. Consider $j < \frac{d}{2}$. Then

$$\begin{aligned}
\sum_{i=j}^{\nu-1} 2^{2j+1}\binom{i+j}{2j} B_{\nu-i} &= 2^{2j}\sum_{i=0}^{n} B_i \binom{i+j+\nu-n}{2j} - 2^{2j+1}\binom{j+\nu}{2j} \\
&= 2^{2j}\sum_{l=0}^{2j}\binom{j+\nu-n}{2j-l}\sum_{i=0}^{n} B_i\binom{i}{l} - 2^{2j+1}\binom{j+\nu}{2j} \\
&= 2^{2j}\sum_{l=0}^{2j}\binom{j+\nu-n}{2j-l} 2^{n-k-l}\binom{n}{l} - 2^{2j+1}\binom{j+\nu}{2j} \\
&= 2^{n-k}\binom{\nu}{j} - 2^{2j+1}\binom{j+\nu}{2j},
\end{aligned}$$

where the second to last equality follows from Theorem 1.3, and so

$$F_j = 2^{2j+1}\binom{j+\nu}{2j} - 2\binom{\nu}{j} \geq 0.$$

Finally, consider $j \geq d_0$. We prove that $F_j \geq 0$ by induction on j. As basis for the induction, we use that $F_{d_0-1} \geq 0$. We observe that $B_{\nu-i} = 0$ for $n - d^\perp < i \leq \nu - 1$. Further

$$\begin{aligned}
\frac{2^{2j+1}\binom{i+j}{2j}}{\binom{\nu}{j}} &= \frac{2(i+j)(i-j+1)}{(2j-1)(\nu-j+1)} \frac{2^{2(j-1)j+1}\binom{i+(j-1)j}{2(j-1)j}}{\binom{\nu}{j-1}} \\
&\leq \frac{2^{2(j-1)j+1}\binom{i+(j-1)j}{2(j-1)j}}{\binom{\nu}{j-1}}
\end{aligned}$$

for $j \ge d_0$ and $j \le i \le \nu - d^\perp$. Hence

$$\begin{aligned}
\frac{F_j}{\binom{\nu}{j}} &= (2^{n-k} - 2) - \sum_{i=j}^{\nu-d^\perp} B_{\nu-i} \frac{2^{2j+1}\binom{i+j}{2j}}{\binom{\nu}{j}} \\
&\ge (2^{n-k} - 2) - \sum_{i=j-1}^{\nu-d^\perp} B_{\nu-i} \frac{2^{2(j-1)j+1}\binom{i+(j-1)j}{2(j-1)j}}{\binom{\nu}{j-1}} \\
&= \frac{F_{j-1}}{\binom{\nu}{j-1}} \ge 0
\end{aligned}$$

by the induction hypothesis. □

Remark 3.2. If

$$\left\lceil \frac{d}{2} \right\rceil \ge \frac{(n - 2d^\perp)^2 + n}{2n},$$

then there are no further values of j to check and similarly for n even. Hence we get the following corollary.

Corollary 3.1. *Let C be an $[n, k, d]$ code where all code words have even weight. If n is odd and*

$$\left\lceil \frac{d}{2} \right\rceil \ge \frac{(n - 2d^\perp)^2 + n}{2n}$$

or n is even and

$$\left\lceil \frac{d}{2} \right\rceil \ge \frac{(n - 2d^\perp)^2 + n - 2}{2n - 2}$$

then

$$P_{\text{ue}}(C, p) \le 2^{k-n} \{1 + (1 - 2p)^n - 2(1 - p)^n\}$$

for all $p \in \left[0, \frac{1}{2}\right]$.

3.2 Binary optimal codes for small dimensions

It is an open question for which n and k there exist $[n, k]$ codes which are optimal for all $p \in \left[0, \frac{1}{2}\right]$. However, for $k \le 4$ such codes do exist for all n; by Corollary 2.12 the same is true for $k \ge n - 4$. Before we can prove this, we must first show a couple of lemmas. The proof is quite long and we divide it into some lemmas. The proof will be based on Corollary 1.7.

Let C be an $[n,k]$ code generated by G whose column count function is m. Then

$$P_{\text{ue}}(C,p) = (1-p)^n \sum_{U \in \mathcal{S}_{k,k-1}} \left(\frac{p}{1-p}\right)^{s(U^c,m)}. \tag{3.2}$$

Lemma 3.1. *If C is optimal for $p \in \left(0, \frac{1}{2}\right)$, then $m(\mathbf{0}) = 0$.*

Proof. Suppose that $m(\mathbf{0}) > 0$. Let $\mathbf{y}$ be an arbitrary non-zero vector in $GF(2)^k$. Define m' by

$$\begin{aligned} m'(\mathbf{0}) &= 0, \\ m'(\mathbf{y}) &= m(\mathbf{y}) + m(\mathbf{0}), \\ m'(\mathbf{x}) &= m(\mathbf{x}) \qquad\qquad \text{for } \mathbf{x} \in GF(2)^k \setminus \{\mathbf{0}, \mathbf{y}\}, \end{aligned}$$

and let C' be a corresponding code. Then

$$s(U^c, m') = \begin{cases} s(U^c, m) + m(\mathbf{0})) & \text{if } \mathbf{y} \in U^c, \\ s(U^c, m) & \text{if } \mathbf{y} \notin U^c. \end{cases}$$

Hence

$$P_{\text{ue}}(C,p) - P_{\text{ue}}(C',p) = (1-p)^n \left\{ 1 - \left(\frac{p}{1-p}\right)^{m(\mathbf{0})} \right\} \sum_{U} \left(\frac{p}{1-p}\right)^{s(U^c,m)} > 0,$$

where the last sum is over all U such that $\mathbf{y} \in U \in \mathcal{S}_{k,k-1}$. □

Let T be an invertible linear transformation on $GF(2)^k$, and let C_T be a code corresponding to $m \circ T$, that is, if m_T corresponds to C_T, then $m_T(\mathbf{x}) = m(T(\mathbf{x}))$.

Lemma 3.2. *We have $P_{\text{ue}}(C_T, p) = P_{\text{ue}}(C, p)$ for all p.*

Proof. We have

$$s(U, m_T) = \sum_{\mathbf{x} \in U} m(T(\mathbf{x})) = \sum_{T(\mathbf{x}) \in TU} m(T(\mathbf{x})) = s(TU, m).$$

Since TU runs through $\mathcal{S}_{k,k-1}$ when U does, we get

$$A(C_T, z) = 1 + \sum_{U \in \mathcal{S}_{k,k-1}} z^{n-s(U,m_T)} = 1 + \sum_{U \in \mathcal{S}_{k,k-1}} z^{n-s(U,m)} = A(C, z).$$

□

Lemma 3.3. *If C is an optimal $[n,k]$ code for $p \in \left(0, \frac{1}{2}\right)$ and $k \le 4$, then*

$$|m(\mathbf{x}) - m(\mathbf{y})| \le 1 \textit{ for } \mathbf{x}, \mathbf{y} \in GF(2)^k \setminus \{\mathbf{0}\}.$$

Proof. We give the proof for $k = 3$. For $k = 1, 2$ the proof is simpler, for $k = 4$ more complicated, but similar. For the details of the proof for $k = 4$ we refer to Kløve (1992).

To simplify the notation, we write

$$m_1 = m(100), m_2 = m(010), \cdots, m_7 = m(111).$$

First we note that there exists a linear transformation on $GF(2)^3$ such that

$$m_T(100) \le m_T(\mathbf{x}) \le m_T(010) \text{ for all } \mathbf{x} \in GF(2)^3 \setminus \{\mathbf{0}\},$$

$$\text{and } m_T(101) \le m_T(011).$$

Hence by Lemma 3.2, we may assume without loss of generality that

$$m_1 \le m_i \le m_2 \text{ for } 1 \le i \le 7,$$

$$m_5 \le m_6.$$

Note that this implies that $m_2 - m_1 \ge m_6 - m_5$. By Corollary 1.7 we get

$$\begin{aligned} A_C(z) = 1 &+ z^{m_1+m_2+m_5+m_6} + z^{m_1+m_2+m_4+m_7} + z^{m_5+m_6+m_4+m_7} \\ &+ z^{m_1+m_5+m_3+m_7} + z^{m_2+m_6+m_3+m_7} \\ &+ z^{m_1+m_6+m_3+m_4} + z^{m_2+m_5+m_3+m_4}. \end{aligned}$$

Suppose that the lemma is not true, that is, C is optimal, but $m_2 - m_1 \ge 2$. By Lemma 3.1, $m_0 = 0$.

Case I, $m_2 - m_1 = m_6 - m_5$. Define $\tilde{m}$ by

$$\tilde{m}_1 = m_1 + 1, \ \tilde{m}_2 = m_2 - 1, \ \tilde{m}_5 = m_5 + 1, \ \tilde{m}_6 = m_6 - 1,$$

and $\tilde{m}_i = m_i$ for $i \in \{0, 3, 4, 7\}$. Then $\tilde{m}_i \ge 0$ for all i. Let $\tilde{C}$ be a corresponding code. Then

$$A_C(z) - A_{\tilde{C}}(z) = z^{m_1+m_3+m_5+m_7}\left(1 - z^2\right)\left(1 - z^{m_2-m_1+m_6-m_5-2}\right) > 0$$

for $z \in (0, 1)$. Hence $P_{\text{ue}}(C, p) > P_{\text{ue}}(\tilde{C}, p)$, contradicting the optimality of C.

Case II, $m_2 - m_1 > m_6 - m_5$. Define $\tilde{m}$ by

$$\tilde{m}_1 = m_1 + 1, \ \tilde{m}_2 = m_2 - 1,$$

and $\tilde{m}_i = m_i$ for $i \in \{0, 3, 4, 5, 6, 7\}$. Again $\tilde{m}_i \ge 0$ for all i. Let $\tilde{C}$ be a corresponding code. Then

$$\begin{aligned} A_C(z) - A_{\tilde{C}}(z) &= z^{m_1+m_3+m_5+m_7}(1 - z)\left(1 - z^{m_2-m_1-1+m_6-m_5}\right) \\ &\quad + z^{m_1+m_3+m_4+m_6}(1 - z)\left(1 - z^{(m_2-m_1)-(m_6-m_5)-1}\right) \\ &> 0 \end{aligned}$$

for $z \in (0, 1)$. Again $P_{\rm ue}(C, p) > P_{\rm ue}(\tilde{C}, p)$, contradicting the optimality of C.

For an $[n, k]$ code C for which $m(\mathbf{x}) > 0$ for all $\mathbf{x} \neq \mathbf{0}$, define C^- to be an $[n - (2^k - 1), k]$ code corresponding to m^- defined by

$$m^-(\mathbf{0}) = 0,$$
$$m^-(\mathbf{x}) = m(\mathbf{x}) - 1 \text{ for } \mathbf{x} \neq \mathbf{0}.$$

Then $(C^-)^* = C$ and

$$p^{2^{k-1}}(1-p)^{2^{k-1}-1} P_{\rm ue}(C^-, p) = P_{\rm ue}(C, p).$$

□

Lemma 3.4. *Let C be an optimal $[n, k]$ code for $p \in \left(0, \frac{1}{2}\right)$ where $k \leq 4$. Then*
(i) C^* is optimal for p,
(ii) if $n \geq 2^k - 1 + k$, then C^- is defined and it is optimal for p.

Proof. Let C_1 be an optimal $[n + 2^k - 1, k]$ code for p. By Lemma 3.3 C_1^- is defined. Hence

$$\begin{aligned} p^{2^{k-1}}(1-p)^{2^{k-1}-1} P_{\rm ue}(C, p) &= P_{\rm ue}(C^*, p) \\ &\geq P_{\rm ue}(C_1, p) = p^{2^{k-1}}(1-p)^{2^{k-1}-1} P_{\rm ue}(C_1^-, p) \end{aligned}$$

and so $P_{\rm ue}(C^*, p) = P_{\rm ue}(C_1, p)$, that is, C^* is optimal for p. This proves (i), and (ii) is similar. □

By Lemma 3.4, for $k \leq 4$ it is sufficient to determine the optimal $[n, k]$ codes for $p \in (0, 1/2)$ where $n < 2^k - 1 + k$. This can be done by a computer search. The search gives the following result.

Theorem 3.3. *If $1 \leq k \leq 4$ and $n \geq k$, then there exists an $[n, k]$ code which is optimal for all $p \in [0, 1/2]$.*

We write $n = r(2^k - 1) + n_0$ where $0 \leq n_0 \leq 2^k - 1$. Then the column count function for an optimal code is given by

$$m_C(\mathbf{x}) = r + m_0(\mathbf{x})$$

and

$$A_C(z) - 1 = z^{r2^{k-1}}\{A_0(z) - 1\}$$

where m_0 and A_0 are given in Tables 3.1–3.3. Note that in the tables, $\mathbf{x}$ is written as a column vector

Table 3.1 $m_0(\mathbf{x})$ and $A_0(z)$ for $k=2$

n_0	$\mathbf{x}$	$A_0(z)$
	101	
	011	
0	000	4
1	100	$2+2z$
2	110	$1+2z+z^2$

Table 3.2 $m_0(\mathbf{x})$ and $A_0(z)$ for $k=3$

n_0	$\mathbf{x}$	$A_0(z)$
	1010101	
	0110011	
	0001111	
0	0000000	8
1	1000000	$4+4z$
2	1100000	$2+4z+2z^2$
3	1101000	$1+3z+3z^2+z^3$
4	1101001	$1+6z^2+z^4$
5	1111100	$1+2z^2+4z^3+z^4$
6	1111110	$1+4z^3+3z^4$

3.3 Modified codes

3.3.1 *Adding/removing a parity bit*

Consider an $[n,k]$ code C containing some code word of odd weight. Adding a parity bit, that is extending each code word $\mathbf{a}$ to $(\mathbf{a}|\sum_{i=1}^{n} a_i)$ gives an $[n+1,k]$ code C^{ex} where all the code words have even weight. The weight distribution function of C^{ex} is given by

$$A_{C^{ex}}(z) = \sum_{i=0}^{\lfloor \frac{n+1}{2} \rfloor} (A_{2i-1} + A_{2i}(C))z^{2i} = \frac{1}{2}\{A_C(z) + A_C(-z)\}.$$

The code C may be good without C^{ex} being good and vice versa. The various combinations of possibilities are given in Figure 3.1 which gives a generator matrix and the weight distribution function of the corresponding codes.

Puncturing the last position of C^{ex} we get C. From the examples above we see that a code may be good without the punctured code being good and vice versa.

Table 3.3 $m_O(\mathbf{x})$ and $A_O(z)$ for $k = 4$

n_O	$\mathbf{x}$	$A_O(z)$
	101010101010101	
	011001100110011	
	000111100001111	
	000000011111111	
0	000000000000000	16
1	100000000000000	$8 + 8z$
2	110000000000000	$4 + 8z + 4z^2$
3	110100000000000	$2 + 6z + 6z^2 + 2z^3$
4	110100010000000	$1 + 4z + 6z^2 + 4z^3 + z^4$
5	110100010000001	$1 + 10z^2 + 5z^4$
6	110100110010000	$1 + 3z^2 + 8z^3 + 3z^4 + z^6$
7	110100110010100	$1 + 7z^3 + 7z^4 + z^7$
8	110100110010110	$1 + 14z^4 + z^8$
9	111110011000011	$1 + 6z^4 + 8z^5 + z^8$
10	111111011000011	$1 + 2z^4 + 8z^5 + 4z^6 + z^8$
11	111111011100110	$1 + 6z^5 + 6z^6 + 2z^7 + z^8$
12	111111011100111	$1 + 12z^6 + 3z^8$
13	111111111111100	$1 + 4z^6 + 8z^7 + 3z^8$
14	111111111111110	$1 + 8z^7 + 7z^8$

3.3.2 *Even-weight subcodes*

Consider an $[n, k]$ code C containing some code word of odd weight. The even-weight $[n, k-1]$ subcode C^e is the set of code words in C of even weight. The weight distribution function of C^e is given by

$$A_{C^e}(z) = \sum_{i=0}^{\lfloor \frac{n}{2} \rfloor} A_{2i}(C) z^{2i}.$$

Figure 3.2 illustrates the various possibilities.

3.4 Binary cyclic redundancy check (CRC) codes

Consider a polynomial

$$g(z) = z^m + g_{m-1} z^{m-1} + \cdots + g_1 z + 1.$$

The $[n = k+m, k]$ code $C_{g,n}$ generated by $g(z)$ is a cyclic redundancy check (CRC) code. In general, the optimal choice of $g(z)$ will depend on k and p. We describe the case $m = 16$ in more detail since a number of such codes are used in practice.

	C^{ex} proper	C^{ex} bad
C proper	$\begin{pmatrix}10\\01\end{pmatrix}$	$\begin{pmatrix}1000001\\0100001\\0011111\end{pmatrix}$
$A_C(z)$:	$1+2z+z^2$	$1+3z^2+3z^5+z^7$
$A_{C^{ex}}(z)$:	$1+3z^2$	$1+3z^2+3z^6+z^8$
C bad	$\begin{pmatrix}100\\010\end{pmatrix}$	$\begin{pmatrix}1000\\0100\end{pmatrix}$
$A_C(z)$:	$1+2z+z^2$	$1+2z+z^2$
$A_{C^{ex}}(z)$:	$1+3z^2$	$1+3z^2$

Fig. 3.1 Codes and extended codes

	C^e proper	C^e bad
C proper	$\begin{pmatrix}1000\\0100\\0011\end{pmatrix}$	$\begin{pmatrix}11000\\01100\\00111\end{pmatrix}$
$A_C(z)$:	$1+2z+2z^2+2z^3+z^4$	$1+3z^2+3z^3+z^5$
$A_{C^e}(z)$:	$1+2z^2+z^4$	$1+3z^2$
C bad	$\begin{pmatrix}1000\\0100\\0010\end{pmatrix}$	$\begin{pmatrix}10000\\01000\\00100\end{pmatrix}$
$A_C(z)$:	$1+3z+3z^2+z^3$	$1+3z+3z^2+z^3$
$A_{C^e}(z)$:	$1+3z^2$	$1+3z^2$

Fig. 3.2 Codes and even-weight subcodes

Examples of CRC codes used in standards are the codes generated by the following polynomials of degree 16:

IEC TC57	$z^{16}+z^{14}+z^{12}+z^{11}+z^9+z^8+z^7+z^4+z+1,$
IEEE WG77.1	$z^{16}+z^{14}+z^{13}+z^{11}+z^{10}+z^9+z^8+z^6+z^5+z+1,$
CCITT X.25	$z^{16}+z^{12}+z^5+1,$
ANSI	$z^{16}+z^{15}+z^2+1,$
IBM-SDLC	$z^{16}+z^{15}+z^{13}+z^7+z^4+z^2+z+1.$

An example for $m=32$ is the ISO 3309 with polynomial

$$z^{32}+z^{26}+z^{23}+z^{22}+z^{16}+z^{12}+z^{11}+z^{10}+z^8+z^7+z^5+z^4+z^2+z+1.$$

In practice, a chosen $g(z)$ will be used for a range of k, and the best choice will depend on the criteria we use.

Methods for efficient computation of the weight distribution of CRC codes has been given by Fujiwara, Kasami, Kitai, and Lin (1985), Miller, Wheal, Stevens, and Lin (1986), Castagnoli, Bräuer, and Herrmann (1993), and Chun and Wolf (1994).

For p sufficiently small, a code with larger minimum distance is better then one with smaller minimum distance. A candidate for an interesting generator polynomial is therefore a polynomial such that the minimum distance of the corresponding $[k+16,k]$ codes is large for a large range of values of k. Castagnoli, Ganz, and Graber (1990) considered the choice of $g(z)$ from this point of view. For example, the polynomial $z^{16}+z^{14}+z^{12}+z^{11}+z^9+z^8+z^7+z^4+z+1$ (used in IEC TC57) generates codes with minimum distance at least 6 for all lengths up to 151, and no other polynomial has this property. In Figure 3.3 we list similar polynomials for other values of $d_{\min}$ (we write only the coefficients in the polynomial).

Based on an exhaustive search of all polynomial with $m=16$, Castagnoli, Ganz, and Graber (1990) gave the optimal choices presented in Figure 3.4. In the table, n_c is the largest length for which $d_{\min}(C_{g,n})>2$. A summary of the main properties of the codes is also given. For more details, we refer to Castagnoli, Ganz, and Graber (1990).

Miller, Wheal, Stevens, and Mezhvinsky (1985) and Miller, Wheal, Stevens, and Lin (1986) considered the polynomials $g(z)=(1+z)p(z)$ where $p(z)$ is irreducible of degree 15. There are 896 such polynomials. They used the criterion that a polynomial $g(z)$ is better if the bound

$$P_{\mathrm{ue}}(C_{g,n},p)\le 2^{-16}\Big\{1-2(1-p)^n+(1-2p)^n\Big\} \tag{3.3}$$

is satisfied over a larger range of values n. By this criterion the best polynomial turned out to be a polynomial that was optimal also by the criterion used by Castagnoli et al., namely the last polynomial in Figure 3.4. For this polynomial, (3.3) is satisfied for all $n\le n_c$.

$d_{\min} \geq$	coeff. of polynomial	length $\leq$
17	11111111111111111	17
12	10101101111101101	18
10	11101001000101111	21
9	11000111101010111	22
8	10001111110110111	31
7	10010011010110101	35
6	10011110101100101	151
5	10101100100110101	257
4	11010001011101011	32767

Fig. 3.3 Generator polynomial which generate codes of a given minimum distance to a given length.

Castagnoli, Bräuer, and Herrmann (1993) and Wolf and Blakeney (1988) have done a similar analysis of polynomials $g(z)$ of degrees 24 and 32.

Wolf and Chun (1994) considered an alternative model for channels with single bursts and the use of CRC codes to detect such burst. In Chun and Wolf (1994) they also describe special hardware to compute the probability of undetected error of CRC codes. Using this hardware they determined polynomials of the form $(z+1)p(z)$, where $p(z)$ is irreducible, such that the corresponding shortened code is good for a large number of test lengths. They gave one polynomial for each degree from 8 through 39. These polynomials are listed in Figures 3.5 and 3.6.

3.5 Particular codes

In this section we consider the error detection of some known classes of binary codes.

3.5.1 *Reed-Muller codes*

The rth order (binary) Reed-Muller code of length 2^m is a $\left[2^m, \sum_{i=0}^{r} \binom{m}{i}\right]$ code. The first order Reed-Muller code is the dual of the extended Hamming code and has weight distribution given in Table 3.4. The code is proper.

coeff. of the polynomial g, properties of the codes $C_{g,n}$	n_c
10011110101100101 $d_{\min} \geq 6$ for $n \leq n_c$, proper for $n \leq n_c$,	151
11111001010011111 $d_{\min} \geq 6$ for $n \leq 130$, $d_{\min} \geq 4$ for $n \leq n_c$, proper for $n \notin \{43, \ldots, 48\} \cup \{189, \ldots, 258\}$, $P_{\rm wc}(C_{g,n}, 0, 0.5)/P_{\rm ue}(C_{g,n}, 0.5) \leq 1.042$ for $n \leq n_c$	258
10101100100110101 $d_{\min} \geq 5$ for $n \leq n_c$, proper for $n \leq n_c$,	257
10000000100011011 $d_{\min} \geq 6$ for $n \leq 115$, $d_{\min} \geq 4$ for $n \leq n_c$, proper when $n \leq 1127$ and $n \notin \{17, \ldots, 28\} \cup \{31, \ldots, 58\}$, $P_{\rm wc}(C_{g,n}, 0, 0.5)/P_{\rm ue}(C_{g,n}, 0.5) \leq 2.11$ for $n \leq n_c$	28658
11010001011101011 $d_{\min} \geq 4$ for $n \leq n_c$, conjectured[†] to be proper for $n \leq n_c$, ([†] Castagnoli et al. verified that $C_{g,n}$ is proper for $n \leq 256$ and a number of larger values of n.)	32767

Fig. 3.4 Generator polynomials and properties of the best CRC codes for $m = 16$.

Table 3.4 Weight distribution of first order Reed-Muller code.

i	0	2^{m-1}	2^m
A_i	1	$2^m - 2$	1

The weight distribution of the second order Reed-Muller code was determined by Sloane and Berlekamp, see MacWilliams and Sloane (1977, p. 443). It is given in Table 3.5. The code is proper for $m \leq 5$, but it is bad for $m \geq 6$. The class of second order Reed-Muller codes is asymptotically

m = degree of $g(z)$	coeff. of the polynomial $p(z)$
8	10001001
9	101100011
10	1000101101
11	11100111001
13	1101110100111
14	10111010011001
15	110100001110111
16	1000011100101001
17	11000101110110111
18	100010010010111011
19	1011101000100010111
20	10010010111000010011
21	110101101110101100011
22	1000011100100110000101
23	10111101110110110100011
24	100010110000111010101011

Fig. 3.5 Generator polynomials $g(z) = (z+1)p(z)$ generating CRC which are good for a range of shortened distances.

bad. In fact, any infinite subset of the set $\{R(r,m) \mid r \geq 2,\, m \geq r+3\}$ is asymptotically bad, see Kløve (1996b).

Table 3.5 Weight distribution of second order Reed-Muller code.

i	A_i
$0,\ 2^m$	1
$2^{m-1} \pm 2^{m-1-h}$	$2^{h(h+1)} \frac{\prod_{i=m-2h+1}^{m}\left(2^i-1\right)}{\prod_{i=1}^{h}\left(2^{2i}-1\right)}$ for $1 \leq h \leq \left\lfloor \frac{m}{2} \right\rfloor$
2^{m-1}	$2^{(m^2+m+2)/2} - 2 - 2\sum_{h=1}^{\lfloor m/2 \rfloor} 2^{h(h+1)} \frac{\prod_{i=m-2h+1}^{m}\left(2^i-1\right)}{\prod_{i=1}^{h}\left(2^{2i}-1\right)}$

Kasami (1971) determined the weight distribution of several subcodes

m = degree of $g(z)$	coeff. of the polynomial $p(z)$
25	1011100100110001111110101
26	11100101101000010110110001
27	100010110010010101110110111
28	1001001010010010101001111001
29	10001011001101101110011101001
30	100110110000010010001000101001
31	1110101100110110111010011111101
32	10100010100010001100010101011001
33	110111011100110101011110100001001
34	1001100010111001001000010010010011
35	10111110010001101111101000010110001
36	110110010011100010101100101110110001
37	1001100001001000011100101110000101101
38	10010000110101010001110010111000000111
39	111011111010001100110101000111100111001

Fig. 3.6 Generator polynomials $g(z) = (z+1)p(z)$ generating CRC which are good for a range of shortened distances.

of the second order Reed-Muller codes.

3.5.2 *Binary BCH codes*

The primitive BCH codes are defined as follows. Let α be a primitive element of $GF(2^m)$. Let $M_j(x)$ be the polynomial of lowest degree over $GF(2)$ having α^j as a root (the *minimal polynomial* of α^j). The t-error correcting BCH code (for short: t-BCH code) is the CRC code generated by the polynomial

$$\operatorname{lcm}\{M_1(x), M_2(x), \ldots, M_{2t}(x)\}.$$

The code has length $n = 2^m - 1$, minimum distance at least $2t + 1$ and dimension at least $n - tm$. The 1-BCH code is the Hamming code. Note that this definition generalizes immediately to q-ary BCH codes.

The weight distribution of the dual code of the binary 2-BCH is given by Tables 3.6 and 3.7. See MacWilliams and Sloane (1977, p. 451f) or Lin and Costello (2004, p. 177f). Leung, Barnes, and Friedman (1979) proved that the binary 2-BCH and the extended binary 2-BCH are both proper.

Table 3.6 Weight distribution of 2-BCH code for m odd.

i	A_i
0	1
$2^{m-1} - 2^{(m-1)/2}$	$\left(2^m - 1\right)\left(2^{m-2} + 2^{(m-3)/2}\right)$
2^{m-1}	$\left(2^m - 1\right)\left(2^{m-1} + 1\right)$
$2^{m-1} + 2^{(m-1)/2}$	$\left(2^m - 1\right)\left(2^{m-2} - 2^{(m-3)/2}\right)$

Table 3.7 Weight distribution of 2-BCH code for m even.

i	A_i
0	1
$2^{m-1} - 2^{m/2}$	$2^{(m-4)/2}\left(2^m - 1\right)\left(2^{(m-2)/2} + 1\right)/3$
$2^{m-1} - 2^{m/2-1}$	$2^{m/2}\left(2^m - 1\right)\left(2^{m/2} + 1\right)/3$
2^{m-1}	$\left(2^m - 1\right)\left(2^{m-2} + 1\right)$
$2^{m-1} + 2^{m/2-1}$	$\frac{1}{3}2^{m/2}\left(2^m - 1\right)\left(2^{m/2} - 1\right)$
$2^{m-1} + 2^{m/2}$	$\frac{1}{3}2^{(m-4)/2}\left(2^m - 1\right)\left(2^{(m-2)/2} - 1\right)$

The binary 3-BCH code is a $[2^m - 1, 2^m - 1 - 3m]$ code whose dual code has the weight distribution given by Tables 3.8 and 3.9, see MacWilliams and Sloane (1977, p. 669) and Lin and Costello (2004, p. 178). Ong and Leung (1991) proved that for m odd the 3-BCH and the corresponding extended code are both proper. For m even, however, Perry (1991) showed that neither the 3-BCH nor the extended 3-BCH are good for $m \geq 6$. However, these classes of codes are asymptotically good.

3.5.3 *Z_4-linear codes*

Let, as usual, Z_4 denote the integers modulo 4. A linear code over Z_4 of length n is a module, that is, a subset C of Z_4^n such that if $\mathbf{u}, \mathbf{v} \in C$, then $a\mathbf{u} + b\mathbf{v} \in C$ for all $a, b \in Z_4$ (the arithmetic is done modulo 4). The dual code $C^\perp$ is defined in the usual way via inner product (modulo 4). Let

Table 3.8 Weight distribution of 3-BCH code for m odd, $m \geq 5$.

i	A_i
0	1
$2^{m-1} - 2^{(m+1)/2}$	$2^{(m-5)/2}\left(2^m - 1\right)\left(2^{m-1} - 1\right)\left(2^{(m-3)/2} + 1\right)/3$
$2^{m-1} - 2^{(m-1)/2}$	$2^{(m-3)/2}\left(2^m - 1\right)\left(5 \cdot 2^{m-1} - 1\right)\left(2^{(m-1)/2} + 1\right)/3$
2^{m-1}	$\left(2^m - 1\right)\left(9 \cdot 2^{2m-4} + 3 \cdot 2^{m-3} + 1\right)$
$2^{m-1} + 2^{(m-1)/2}$	$2^{(m-3)/2}\left(2^m - 1\right)\left(5 \cdot 2^{m-1} - 1\right)\left(2^{(m-1)/2} - 1\right)/3$
$2^{m-1} + 2^{(m+1)/2}$	$2^{(m-5)/2}\left(2^m - 1\right)\left(2^{m-1} - 1\right)\left(2^{(m-3)/2} - 1\right)/3$

Table 3.9 Weight distribution of the 3-BCH code for m even, $m \geq 6$.

i	A_i
0	1
$2^{m-1} - 2^{(m+2)/2}$	$\left(2^m - 1\right)\left(2^m - 4\right)\left(2^{m-1} + 2^{(m+2)/2}\right)/960$
$2^{m-1} - 2^{m/2}$	$7\left(2^m - 1\right)2^m\left(2^{m-1} + 2^{m/2}\right)/48$
$2^{m-1} - 2^{(m-2)/2}$	$\left(2^m - 1\right)\left(2^{m-1} + 2^{(m-2)/2}\right)\left(6 \cdot 2^m + 16\right)/15$
2^{m-1}	$\left(2^m - 1\right)\left(29 \cdot 2^{2m} - 2^{m+2} + 64\right)/64$
$2^{m-1} + 2^{(m-2)/2}$	$\left(2^m - 1\right)\left(2^{m-1} - 2^{(m-2)/2}\right)\left(6 \cdot 2^m + 16\right)/15$
$2^{m-1} + 2^{m/2}$	$7\left(2^m - 1\right)2^m\left(2^{m-1} - 2^{m/2}\right)/48$
$2^{m-1} + 2^{(m+2)/2}$	$\left(2^m - 1\right)\left(2^m - 4\right)\left(2^{m-1} - 2^{(m+2)/2}\right)/960$

$\phi : Z_4 \to GF(2)^2$ be defined by

$$\phi(0) = (00), \quad \phi(1) = (01), \quad \phi(2) = (11), \quad \phi(3) = (10),$$

and $\phi : Z_4^n \to GF(2)^{2n}$ by

$$\phi(v_1, v_2, \ldots, v_n) = (\phi(v_1)|\phi(v_2)|\ldots|\phi(v_n)).$$

Finally, for a linear code C over Z_4 of length n we define the binary code $\phi(C)$ of length $2n$ by

$$\phi(C) = \{\phi(\mathbf{v}) \mid \mathbf{v} \in C\}.$$

Note that $\phi(C)$ is not in general linear; such codes have been termed Z_4-linear. For a more detailed description of these concepts we refer to the paper by Hammons, Kumar, *et al.* (1994). In particular they prove the following two results which are important in our context:

- A Z_4-linear code is distance invariant, in particular,

$$A_{\phi(C)}(z) = A^{\mathrm{w}}_{\phi(C)}(z),$$

- $A_{\phi(C^\perp)}(z) = A^{\mathrm{MW}}_{\phi(C)}(z)$.

Note that both $\phi(C)$ and $\phi(C^\perp)$ may be non-linear and are not dual in the usual sense.

One class of codes which can be obtained this way is the *Kerdock codes* $\mathcal{K}(m)$ which are non-linear $(2^m, 2^{2m})$ codes. The distance distribution is given in Tables 3.10, see MacWilliams and Sloane (1977, p. 456) or Hammons, Kumar, *et al.* (1994).

Table 3.10 Distance distribution, Kerdock codes.

i	A_i for m even	A_i for m odd
$0,\ 2^m$	1	1
$2^{m-1} \pm 2^{(m-2)/2}$	$2^m\left(2^{m-1}-1\right)$	-
$2^{m-1} \pm 2^{(m-1)/2}$	-	$2^{m-1}\left(2^{m-1}-1\right)$
2^{m-1}	$2^{m+1}-2$	$2^m\left(2^{m-1}+1\right)-2$

The *Preparata codes* are another class of binary Z_4-linear $(2^m, 2^{2^m-2m})$ codes. The distance distribution of the Preparata code is the MacWilliams transform of the distance distribution of the corresponding Kerdock code. Dodunekova, Dodunekov and Nikolova (2004a) showed that both the Kerdock codes and the Preparata codes are proper.

The *Delsarte-Goethals* $\mathcal{DG}(m,d)$ *codes*, where $m \geq 6$ is even, are non-linear $(2^m, 2^{(m-1)(\lfloor m/2 \rfloor - d+1)+m+1})$ codes. In particular $\mathcal{DG}(m, m/2) = \mathcal{K}(m)$. The distance distribution of the $\mathcal{DG}(m, m/2-1)$ codes is given in Table 3.11, (see MacWilliams and Sloane (1977, p. 477)).

Table 3.11 Distance distribution, $\mathcal{DG}(m, m/2-1)$ codes.

i	A_i
$0, 2^m$	1
$2^{m-1} \pm 2^{m/2}$	$2^{m-2}\left(2^{m-1}-1\right)\left(2^m-1\right)/3$
$2^{m-1} \pm 2^{m/2-1}$	$2^m\left(2^{m-1}-1\right)\left(2^{m-1}+4\right)/3$
2^{m-1}	$2\left(2^m-1\right)\left(2^{2m-3}-2^{m-2}+1\right)$

3.5.4 Self-complementary codes

A binary linear code C is self-complementary if the complement of any code word is again a code word, that is, if $\mathbf{c} \in C$, then $\mathbf{c}+\mathbf{1} \in C$, where $\mathbf{1}$ is the all-one vector. For example, the codes considered in Section 3.1 are duals of self-complementary codes. For a self-complementary $[n,k]$ code C, the weight distribution is symmetric, that is, $A_i(C) = A_{n-i}(C)$ for all i. For the study of such codes, we start with a lemma.

Lemma 3.5. *Let*

$$\frac{n-\sqrt{n}}{2} \leq i < \frac{1}{2}.$$

Then the function

$$f(p) = p^i(1-p)^{n-i} + p^{n-i}(1-p)^i$$

is increasing on $[0,1/2]$.

Proof. We have

$$\begin{aligned} f'(p) &= ip^{i-1}(1-p)^{n-i} - (n-i)p^i(1-p)^{n-i-1} \\ &\quad +(n-i)p^{n-i-1}(1-p)^i - ip^{n-i}(1-p)^{i-1} \\ &= p^{n-i-1}(1-p)^{i-1}(n-i-np)g(p), \end{aligned}$$

where

$$g(p) = 1 - \left(\frac{1-p}{p}\right)^{n-2i} \frac{np-i}{n-i-np}.$$

Further

$$\begin{aligned} g'(p) &= \left(\frac{1-p}{p}\right)^{n-2i-1} \frac{1}{p^2} \frac{(n-2i)}{(n-i-np)^2} \Big\{np(1-p) - (np-i)(n-i-np)\Big\} \\ &= \left(\frac{1-p}{p}\right)^{n-2i-1} \frac{1}{p^2} \frac{(n-2i)}{(n-i-np)^2} \\ &\quad \cdot\Big\{(n^2-n)\Big(p-\frac{1}{2}\Big)^2 - \frac{n^2-n}{4} + i(n-i)\Big\}. \end{aligned}$$

We see that $g'(p)$ has its minimum for $p = 1/2$. Moreover, this minimum is non-negative since

$$i(n-i) - \frac{n^2-n}{4} \geq \frac{n-\sqrt{n}}{2}\Big(n - \frac{n-\sqrt{n}}{2}\Big) - \frac{n^2-n}{4} = 0.$$

Hence $f'(p) \geq 0$ for all $p \in [0,1/2]$, that is, $f(p)$ is increasing. $\square$

From the lemma we immediately get the following result.

Theorem 3.4. *If C is an (n, M, d) code with symmetric distance distribution and $d \geq \frac{n-\sqrt{n}}{2}$, then C is proper.*

Remark 3.3. If C is a self-complementary code (n, M, d) and $d > \frac{n-\sqrt{n}}{2}$, that is, $n - (n-2d)^2 > 0$, then $M \leq \frac{8d(n-d)}{n-(n-2d)^2}$; this is known as the Grey-Rankin bound. Further, the condition of Corollary 3.1 is also satisfied and so (3.1) is satisfied.

3.5.5 *Self-dual codes*

A linear code C is self-dual if $C^{\perp} = C$. For a self-dual code $k = n - k$, that is $n = 2k$. The binary self-dual codes of dimension up to 16 has been classified, for some larger values of k partial classification has been done. The weight distribution has been determined for many self-dual codes. In particular, the classification and weight distributions of all binary self-dual $[32, 16]$ codes was determined by Bilous and van Rees (2002). An excellent overview of self-dual code is given by Rains and Sloane in Pless and Huffman (1998) pp. 177–294.

We note that $\mathbf{c} \cdot \mathbf{c} = 0$ if and only if $\mathbf{c}$ has even weight. Hence, all the code words of a self-dual C code have even weight. In particular, this implies that the all-one vector is contained in $C^{\perp} = C$, and so C is in particular self-complementary and $A_{n-i}(C) = A_i(C)$ for all i.

Example 3.1. Following Perry and Fossorier (2006b), we describe the error detecting capability of the self-dual $[32, 16]$ codes. From Theorem 2.10, we see that a necessary condition for a self-dual $[32, 16]$ code to be good is $A_2 = 0$ and $A_4 \leq 2$. From the tables in Bilous and van Rees (2002) we see that there are exactly 29 possible weight distributions that satisfy these conditions. These weight distributions fall into four classes given in Table 3.12.

The first class is proper for all b.

The second class is proper for $b \leq 7$, bad for $b = 8, 9$.

The third class is good but not proper for $b \leq 3$, bad for $b \geq 4$.

The forth class is proper for all b.

For those codes that are proper, this can in all cases be shown using the sufficient condition in Theorem 2.19. The code with $b = 3$ in the third class is an interesting example of a good code with two local maxima in the interval $(0, 1/2)$, namely for $p \approx 0.1628$ and $p \approx 0.4109$. Also the (bad)

Table 3.12 Classes of weight distributions for self-dual $[32, 16]$ codes

Class:	1	2	3	4
A_4	0	1	2	b
A_6	$4b$	$4b$	$4b$	0
A_8	$364 - 8b$	$374 - 8b$	$384 - 8b$	$620 + 10b$
A_{10}	$2048 - 12b$	$2048 - 12b$	$2048 - 12b$	0
A_{12}	$6720 + 32b$	$6771 + 32b$	$6622 + 32b$	$13888 - 49b$
A_{14}	$14336 + 8b$	$14336 + 8b$	$14336 + 8b$	0
A_{16}	$18598 - 48b$	$18674 - 48b$	$18750 - 48b$	$36518 + 76b$
range:	$b = 0$ and $2 \leq b \leq 8$	$2 \leq b \leq 9$	$b = 0$ and $2 \leq b \leq 10$	$b = 0, 1, 2$

code with $b = 4$ in this class has two maxima.

3.6 Binary constant weight codes

For given n and m, let Ω_n^m denote the set of all binary vectors of length n and weight m. A (binary) constant weight code is some subset of Ω_n^m. The distance between two code words in a binary constant weight code is clearly even. If the minimum distance is 2δ, we call the code an $(n, M, 2\delta, m)$ code. Note that these codes are not linear (in particular, the zero vector is not a code word).

In this section we will first give some results for the codes Ω_n^m and next results for constant weight codes in general, that is, subcodes of Ω_n^m. The code Ω_n^{n-m} is essentially the same code as Ω_n^m (we have only interchanged the zeros and the ones). In particular, $P_{\text{ue}}(\Omega_n^{n-m}, p) = P_{\text{ue}}(\Omega_n^m, p)$. Therefore, we will assume that $m \leq n - m$, that is $m \leq \lfloor n/2 \rfloor$. In this section we mainly quote known results without proofs. However, we give references for the results.

3.6.1 *The codes* Ω_n^m

Theorem 3.5. *Let* $0 \leq m \leq \lfloor n/2 \rfloor$. *For all* i, $0 \leq j \leq m$ *we have*

$$A_{2j}(\Omega_n^m) = \binom{m}{j}\binom{n-m}{j}.$$

For all other i *we have* $A_i(\Omega_n^m) = 0$.

Proof. For any code words in Ω_n^m, we obtain a code word at distance $2j$ exactly when j of the m ones are changed to zeros and j of the $n - m$ zeros are changed to ones. The number of ways to choose the j ones is $\binom{m}{j}$ and the number of ways to choose the j zeros is $\binom{n-m}{j}$. □

Theorem 3.6. *a) For $n \le 4$, all Ω_n^m codes are proper.*
b) For $5 \le n \le 8$, the codes $\Omega_n^{\lfloor n/2 \rfloor}$ are proper.
c) For all other n and $m \le \lfloor n/2 \rfloor$, the codes Ω_n^m are bad.

The threshold was defined in (2.3). In particular, $\theta(\Omega_n^m)$ is the smallest root in the interval $(0, 1/2]$ of the equation $P_{ud}(\Omega_n^m, p) = P_{ud}(\Omega_n^m, 1/2)$. For $p \le \theta(C)$, the bound $P_{ud}(\Omega_n^m, p) \le P_{ud}(\Omega_n^m, 1/2)$ is valid.

Theorem 3.7. *Let*

$$\psi = \psi(n, m) = \left(\frac{\binom{n}{m} - 1}{2^n m(n-m)} \right)^{1/2}.$$

i) For all n and m such that $1 \le m < n$ we have

$$\psi(n, m) < \left(\frac{32}{\pi n^5} \right)^{1/4}.$$

ii) For all sufficiently large n and all m such that $1 \le m < n$ we have

$$\omega(n, m) \le \theta(\Omega_n^m) \le \omega(n, m) + n^2 \psi(n, m)^3$$

where

$$\omega(n, m) = \psi(n, m) + \frac{(n-2)}{2} \psi(n, m)^2.$$

Corollary 3.2. *We have*

$$\lim_{n \to \infty} 2^{n/2} \theta(\Omega_n^1) = 1.$$

Corollary 3.3. *If $0 < \lambda \le 1/2$, then*

$$\lim_{n \to \infty} n^{5/4} 2^{n\left(1 - H_2(\lambda)\right)/2} \theta(\Omega_n^{\lambda n}) = \frac{1}{\left(2\pi \lambda^3 (1-\lambda)^3\right)^{1/4}},$$

where $H_2(\lambda)$ is the (binary) entropy function.

Corollary 3.4. *We have*

$$\lim_{n \to \infty} n^{5/4} \theta(\Omega_n^{n/2}) = \left(\frac{32}{\pi} \right)^{1/4}.$$

Let $P_{ue}(n, M, w, p)$ denote the minimum value of $P_{ue}(C, p)$ over all binary $(n, M, 2, w)$ constant weight codes. A binary $(n, M, 2, w)$ code C is called optimal (error detecting) for p if $P_{ue}(C, p) = P_{ue}(n, M, w, p)$.

3.6.2 *An upper bound*

Let $\mathcal{C}(n, M, w)$ be the set consisting of all binary $(n, M, 2, w)$ constant weight codes. The mean probability of undetected error for the codes in $\mathcal{C}(n, M, w)$ is given by

$$\bar{P}_{ue}(n, M, w, p) = \frac{1}{\#\mathcal{C}(n, M, w)} \sum_{C \in \mathcal{C}(n,M,w)} P_{ue}(C, p).$$

Theorem 3.8.

$$\bar{P}_{ue}(n, M, w, p) = \frac{(M-1)\binom{n}{w}}{[\binom{n}{w} - M + 1][\binom{n}{w} - M + 2]} \sum_{i=1}^{w} \binom{w}{i}\binom{n-w}{i} p^{2i}(1-p)^{n-2i}.$$

Corollary 3.5.

$$P_{ue}(n, M, w, p) \leq \frac{(M-1)\binom{n}{w}}{[\binom{n}{w} - M + 1][\binom{n}{w} - M + 2]} \cdot \sum_{i=1}^{w} \binom{w}{i}\binom{n-w}{i} p^{2i}(1-p)^{n-2i}.$$

3.6.3 *Lower bounds*

Theorem 3.9. *If C is a binary $(n, M, 2, w)$ constant weight code, then*

$$P_{ue}(C, p) \geq (M-1)(1-p)^n \left(\frac{p}{1-p}\right)^{\frac{2w(n-w)M}{n(M-1)}}.$$

Theorem 3.10. *Let C be a binary $(n, M, 2, w)$ constant weight code, then*

$$P_{ue}(C, p) \geq \frac{M}{\binom{n}{w}} \Big[\sum_{i=1}^{w} \binom{w}{i}\binom{n-w}{i} p^{2i}(1-p)^{n-2i}\Big] - \Big[1 - \frac{M}{\binom{n}{w}}\Big](1-p)^n.$$

There are a couple of lower bounds which are analogous to Theorem 2.44.

Theorem 3.11. *Let C be a binary $(n, M, 2, w)$ constant weight code. Then*

$$P_{ue}(C, p) \geq \sum_{l=1}^{n} \max\{0, F_l(n, M, w)\} p^l (1-2p)^{n-l}$$

where

$$F_l(n, M, w) = M \sum_{t=\max\{0, w-l\}}^{\min\{w, n-l\}} \frac{\binom{w}{t}^2 \binom{n-w}{n-l-t}^2}{\binom{n-l}{t}\binom{n}{l}} - \binom{n}{l}.$$

Theorem 3.12. *Let C be a binary $(n, M, 2, w)$ constant weight code such that*

$$\frac{\binom{n}{w}}{\binom{n-t}{w-t}} \le M < \frac{\binom{n}{w}}{\binom{n-t-1}{w-t-1}}$$

for some t, where $1 \le t < w$. Then

$$P_{ue}(C,p) \ge (1-p)^{n-2w} \sum_{l=w-t}^{t} \left[\frac{M\binom{n-w+l}{l}}{\binom{n}{w}} - 1 \right] \binom{w}{l} p^{2l}(1-2p)^{n-l}.$$

If D is a $t-(v,k,\lambda)$ block design, the rows of an incidence matrix for D form a constant weight code C_D of length v, size $\lambda\frac{\binom{v}{t}}{\binom{k}{t}}$, and weight k.

Theorem 3.13. *Let C_D be a binary (v, M, d, k) constant weight code obtained from a $t-(v,k,\lambda)$ block design such that $\lambda < (n-t)/(w-t)$ and $d \ge 2(w-t)$. Then C_D is an optimal constant weight code for all $p \in [0, 1/2]$ and*

$$P_{ue}(C,p) = (1-p)^{n-2w} \sum_{l=w-t}^{t} \left[\frac{\lambda\binom{n-w+l}{l}}{\binom{n-t}{w-t}} - 1 \right] \binom{w}{l} p^{2l}(1-2p)^{n-l}.$$

3.7 Comments and references

3.1. Theorems 3.1 and 3.2 are from Kasami, Kløve, and Lin (1983).

3.2. The results are taken from Kløve (1992).

3.3. The results and examples are taken from Kløve and Korzhik (1995).

3.4. The referred standard CRC codes are mainly collected from the internet.

3.5. Lemma 3.5 and Theorem 3.4 are due to Dodunekova, Dodunekov and Nikolova (2004a) (actually, they gave a more general version of the lemma).

3.6. Theorem 3.6 was shown by Wang, Yang, and Zhang, see Wang (1987), Wang (1989), Wang (1992), Wang and Yang (1994), Wang and Zhang (1995), Yang (1989). A simpler proof was given by Fu, Kløve, and Xia (2000a).

Fu, Kløve, and Xia (2000a) proved Theorem 3.7 and its corollaries.

Theorems 3.9, 3.10, and 3.11 are due to Fu, Kløve, and Wei (2003).

Theorems 3.12 and 3.13 are due to Xia, Fu, and Ling (2006a).

In addition to the result listed, a number of asymptotic bounds were given by Fu, Kløve, and Wei (2003) and Xia, Fu, and Ling (2006a).

Chapter 4

Error detecting codes for asymmetric and other channels

4.1 Asymmetric channels

Let the alphabet be Z_q with the ordering $0 < 1 < 2 \cdots < q-1$. A channel is called asymmetric if any transmitted symbol a is received as $b \le a$. For example, for $q = 2$, a 0 is always received correctly while a 1 may be received as 0 or 1. The binary channel where

$$\pi(0|0) = 1, \quad \pi(1|0) = 0, \quad \pi(0|1) = p, \quad \pi(1|1) = 1 - p$$

is known as the Z-channel.

An asymmetric channel is called complete if $\pi(b|a) > 0$ for all $b \le a$.

For general q, the two main complete channels considered are the one where each error is equally probable and the one where the errors are given weight proportional to $a - b$. For both channels $\pi(b|a) = 0$ if $b > a$ and $\pi(0|0) = 1$. For $a > 0$, the first channel is defined by

$$\pi(b|a) = \begin{cases} 1 - p \text{ if } b = a, \\ p/a \quad \text{if } b < a. \end{cases}$$

and the second channel is defined by

$$\pi(b|a) = \begin{cases} 1 - p & \text{if } b = a, \\ \frac{a-b}{a(a-1)/2} p & \text{if } b < a. \end{cases}$$

We note that for $q = 2$, these two channels are the same, namely the Z-channel. We first give some results for the Z-channel.

4.1.1 *The Z-channel*

Let $\mathbf{x}$ be sent and $\mathbf{y}$. By definition, $P(\mathbf{y}|\mathbf{x}) = 0$ if $\mathbf{y} \not\le \mathbf{x}$. If $\mathbf{y} \le \mathbf{x}$, then

$$P(\mathbf{y}|\mathbf{x}) = p^{d_H(\mathbf{x},\mathbf{y})}(1-p)^{w_H(\mathbf{x})-d_H(\mathbf{x},\mathbf{y})} = p^{w_H(\mathbf{x})-w_H(\mathbf{y})}(1-p)^{w_H(\mathbf{y})}.$$

Therefore

$$P_{\rm ue}(C, Z_p) = \frac{1}{\#C} \sum_{\mathbf{x}\in C} \sum_{\substack{\mathbf{y}\in C\\ \mathbf{y}<\mathbf{x}}} p^{w_H(\mathbf{x})-w_H(\mathbf{y})}(1-p)^{w_H(\mathbf{y})}. \tag{4.1}$$

It is interesting to compare $P_{\rm ue}(C, Z_p)$ and $P_{\rm ue}(C, BSC_p)$ for some codes. If $\mathbf{y} \not< \mathbf{x}$ for all $\mathbf{y}, \mathbf{x} \in C$, then clearly $P_{\rm ue}(C, Z_p) = 0$, whereas $P_{\rm ue}(C, BSC_p) > 0$ (if $\#C \geq 2$ and $p \in (0,1)$). For other codes, the ratio $P_{\rm ue}(C, Z_p)/P_{\rm ue}(C, BSC_p)$ may be large or small, depending on the code and p.

Example 4.1. As a simple example, consider the $[2^k-1, k]$ simplex code S_k. This is a linear code where all non-zero code words have Hamming weight 2^{k-1}. Hence $\mathbf{y} < \mathbf{x}$ for $\mathbf{y}, \mathbf{x} \in S_k$ if and only if $\mathbf{y} = \mathbf{0}$ and $\mathbf{x} \neq \mathbf{0}$. Therefore

$$P_{\rm ue}(S_k, Z_p) = \frac{2^k-1}{2^k} p^{2^{k-1}}$$

and

$$P_{\rm ue}(S_k, BSC_p) = (2^k-1)p^{2^{k-1}}(1-p)^{2^{k-1}-1}.$$

Hence

$$\frac{P_{\rm ue}(S_k, Z_p)}{P_{\rm ue}(S_k, BSC_p)} \to \frac{1}{2^k} \quad \text{when} \quad p \to 0,$$

and

$$\frac{P_{\rm ue}(S_k, Z_p)}{P_{\rm ue}(S_k, BSC_p)} \to \infty \quad \text{when} \quad p \to 1.$$

Also, for any $p < 1$ we have

$$\frac{P_{\rm ue}(S_k, Z_p)}{P_{\rm ue}(S_k, BSC_p)} \to \infty \quad \text{when} \quad k \to \infty.$$

A code C is called *perfect for error detection* on the Z-channel if $P_{\rm ue}(C, Z_p) = 0$ for all p. By (4.1), C is perfect if and only if $\mathbf{y} \not< \mathbf{x}$ for all $\mathbf{y}, \mathbf{x} \in C$, $\mathbf{y} \neq \mathbf{x}$. The largest perfect C of length n is obtained by taking all vectors of weight $\lfloor \frac{n}{2} \rfloor$.

A perfect systematic $(k + \lceil \log_2 k \rceil, 2^k)$ code can be constructed as follows:

$$C_k = \{(\mathbf{x}|r(\mathbf{x})) \mid \mathbf{x} \in Z_2^k\}$$

where $r(\mathbf{x}) \in Z_2^{\lceil \log_2 k \rceil}$ is obtained by first taking the binary expansion of $w_H(\mathbf{x})$ and then taking the binary complement (changing 0 to 1 and 1 to 0). E.g.

$$C_3 = \{00011, 00110, 01010, 01101, 10010, 10101, 11001, 11100\}.$$

We note that C_k has 2^k code words whereas the non-systematic perfect code of the same length given above has

$$\binom{k + \lceil \log_2 k \rceil}{\left\lfloor \frac{k+\log_2 k}{2} \right\rfloor} \approx \sqrt{\frac{2k}{\pi}} 2^k$$

code words.

One general construction of (non-perfect) error detecting codes for the Z-channel is given in the next theorem.

Theorem 4.1. *Let a_i be integers for $1 \le i \le k$ such that*

$$0 \le a_1 < a_2 < \cdots < a_k \le n.$$

Let

$$C = \{\mathbf{x} \in Z_2^n \mid w_H(\mathbf{x}) = a_i \text{ for some } i\}.$$

Then

$$P_{\mathrm{ue}}(C, Z_p) = \frac{1}{\sum_{i=1}^{k} \binom{n}{a_i}} \sum_{i=2}^{k} \sum_{j=1}^{i-1} \binom{n}{a_i} \binom{a_i}{a_j} p^{a_i - a_j} (1-p)^{a_j}.$$

Proof. There are $\binom{n}{a_i}$ vectors $\mathbf{x} \in C$ of Hamming weight a_i. For each of these there are $\binom{a_i}{a_j}$ vectors $\mathbf{y} \in C$ of Hamming weight a_j such that $\mathbf{y} < \mathbf{x}$. □

Example 4.2. Let $a_i = 2i - 1$ for $i = 1, 2, \ldots, k$ where $k = \lceil n/2 \rceil$, that is, we take all vectors of odd weight. The corresponding code C_{o} detects all single asymmetric errors (and many others), and from Theorem 4.1 we get after simplifications:

$$P_{\mathrm{ue}}(C_{\mathrm{o}}, Z_p) = \frac{1}{2} + \frac{1}{2}(1-p)^n - \left(1 - \frac{p}{2}\right)^n - \left(\frac{1}{2} - \frac{1}{2^n}\right) p^n.$$

Another possibility is to let $a_i = 2i$ for $i = 0, 1, \ldots, k$ where $k = \lfloor n/2 \rfloor$, that is, we take all vectors of even weight. For the corresponding code C_{e} we get

$$P_{\mathrm{ue}}(C_{\mathrm{e}}, Z_p) = \frac{1}{2} + \frac{1}{2}(1-p)^n - \left(1 - \frac{p}{2}\right)^n + \left(\frac{1}{2} - \frac{1}{2^n}\right) p^n.$$

In particular, $\#C_{\mathrm{o}} = \#C_{\mathrm{e}}$ and $P_{\mathrm{ue}}(C_{\mathrm{o}}, Z_p) < P_{\mathrm{ue}}(C_{\mathrm{e}}, Z_p)$ for all $n > 1$, $p \ne 0$.

4.1.2 *Codes for the q-ary asymmetric channel*

For $\mathbf{x} = (x_0, x_1, \ldots, x_{k-1}) \in Z_q^k$, let

$$w_q(\mathbf{x}) = \sum_{i=0}^{k-1} x_i, \text{ the } \textit{weight} \text{ of } \mathbf{x},$$

$$u_q(\mathbf{x}) = \sum_{i=0}^{k-1} (q-1-x_i), \text{ the } \textit{coweight} \text{ of } \mathbf{x}.$$

Clearly, $w_q(\mathbf{x}) + u_q(\mathbf{x}) = k(q-1)$.

For non-negative integers a and s, where $a < q^s$, let

$$\langle a \rangle_s = (a_0, a_1, \ldots, a_{s-1}) \in Z_q^s$$

where

$$a = \sum_{i=0}^{s-1} a_i q^i, \; a_i \in Z_q.$$

Define

$$w_q(a) = w_q(\langle a \rangle_s) = \sum_{i=0}^{s-1} a_i.$$

For example, if $q = 5$ and $k = 6$, then

$$w_5((1,3,1,0,2,4)) = 11 \text{ and } u_5((1,3,1,0,2,4)) = 13.$$

Further, $\langle 33 \rangle_4 = (3,1,1,0)$ since $33 = 3 + 1 \cdot 5 + 1 \cdot 5^2 + 0 \cdot 5^3$, and $w_4(33) = 5$.

For integers a and n, let $[a]_n$ denote the (least non-negative) residue of a modulo n.

For integers $m \geq 0$ and $n \geq 0$, let $S(m,n)$ denote the number of arrays in Z_q^m of weight n.

Theorem 4.2. *For a complete q-ary asymmetric channel, a maximal perfect code of length m is*

$$\left\{ \mathbf{x} \in Z_q^m \;\middle|\; w_q(\mathbf{x}) = \left\lfloor \frac{m(q-1)}{2} \right\rfloor \right\}.$$

The size of this code is $S\left(m, \left\lfloor \frac{m(q-1)}{2} \right\rfloor\right)$.

Generalized Bose-Lin codes (GBL codes) is a class of systematic codes. They are determined by four (integral) parameters, $q \geq 2$ (symbol alphabet size), k (number of information symbols), r (number of check symbols), and

ω where $0 \le \omega \le r$. We use the notations $\rho = r - \omega$, $\sigma = S(\omega, \lfloor \omega(q-1)/2 \rfloor)$, $\theta = q^{\rho}$, and $\mu = \sigma\theta$. Finally, let

$$\{\mathbf{b}_{\omega,0}, \mathbf{b}_{\omega,1}, \ldots, \mathbf{b}_{\omega,\sigma-1}\},$$

the set of vectors in Z_q^{ω} of weight $\left\lfloor \frac{\omega(q-1)}{2} \right\rfloor$.

A code word is a vector in Z_q^{k+r}. It consists of an information part $\mathbf{x}$ with k information symbols concatenated by a check part $c(\mathbf{x})$ with r check symbols. Let $u = u_q(\mathbf{x})$. The check part, which depends only on $[u]_{\mu}$, will be determined as follows. Let $\alpha = \lfloor [u]_{\mu}/\theta \rfloor$. Then $0 \le \alpha < \sigma$ and

$$[u]_{\mu} = \alpha\theta + [u]_{\theta}.$$

The *check part* is defined by $c(\mathbf{x}) = (c_1(\mathbf{x})|c_2(\mathbf{x}))$, where

$$c_1(\mathbf{x}) = \mathbf{b}_{\omega,\alpha} \text{ and } c_2(\mathbf{x}) = \langle [u]_{\theta} \rangle_{\rho}.$$

As usual, an undetectable error occurs if a code word is sent and the received array is another code word. We characterize the undetectable errors. Assume that $(\mathbf{x}|c_1(\mathbf{x})|c_2(\mathbf{x}))$ is the sent code word and $(\mathbf{y}|c_1(\mathbf{y})|c_2(\mathbf{y}))$ is the received code word, where $\mathbf{y} \ne \mathbf{x}$. This is possible if and only if

$$\mathbf{y} \subset \mathbf{x}, \tag{4.2}$$

$$c_1(\mathbf{y}) = c_1(\mathbf{x}) \tag{4.3}$$

since $w_q(c_1(\mathbf{y})) = w_q(c_1(\mathbf{x}))$, and

$$c_2(\mathbf{y}) \subseteq c_2(\mathbf{x}). \tag{4.4}$$

Suppose that (4.2)–(4.4) are satisfied, and let

$$u = u_q(\mathbf{x}) \tag{4.5}$$

and

$$j\mu - \lambda = u_q(\mathbf{y}) - u_q(\mathbf{x}), \tag{4.6}$$

where $j \ge 1$ and $0 \le \lambda < \mu$. Note that $\lambda = [u_q(\mathbf{x}) - u_q(\mathbf{y})]_{\mu}$. Let

$$[u]_{\mu} = \alpha\theta + [u]_{\theta} \text{ and } [u-\lambda]_{\mu} = \beta\theta + [u-\lambda]_{\theta}.$$

Since $[u-\lambda]_{\mu} = [u + (j\mu - \lambda)]_{\mu}$, we get

$$\begin{aligned} c_1(\mathbf{x}) &= \mathbf{b}_{\omega,\alpha}\ c_2(\mathbf{x}) = \langle [u]_{\theta} \rangle_{\rho}, \\ c_1(\mathbf{y}) &= \mathbf{b}_{\omega,\beta}\ c_2(\mathbf{y}) = \langle [u-\lambda]_{\theta} \rangle_{\rho}. \end{aligned}$$

Hence, (4.3) is satisfied if and only if

$$\beta = \alpha, \tag{4.7}$$

and (4.4) is satisfied if and only if

$$\langle [u-\lambda]_\theta \rangle_\rho \subseteq \langle [u]_\theta \rangle_\rho. \tag{4.8}$$

We observe that if $\lambda \geq \theta$, then $\alpha \neq \beta$, and if $[u]_\theta < \lambda < \theta$, then (4.8) cannot be satisfied. On the other hand, if $\lambda \leq [u]_\theta$, then (4.7) is satisfied; further (4.8) is satisfied exactly when $\langle \lambda \rangle_\rho \subseteq \langle [u]_\theta \rangle_\rho$. We also note that $\langle \lambda \rangle_\rho \subseteq \langle [u]_\theta \rangle_\rho$ implies that $\lambda \leq [u]_\theta$. Hence, we have proved the following result.

Theorem 4.3. *The code word* $(\mathbf{x}|c(\mathbf{x}))$ *can be transformed to the code word* $(\mathbf{y}|c(\mathbf{y}) \neq \mathbf{x}|c(\mathbf{x}))$ *by transmission over the q-ASC if and only if*

$$\mathbf{y} \subset \mathbf{x} \text{ and } \langle \lambda \rangle_\rho \subseteq \langle [u_q(\mathbf{x})]_\theta \rangle_\rho,$$

where $\lambda = [u_q(\mathbf{x}) - u_q(\mathbf{y})]_\mu$.

We now consider the minimal weight of an undetectable error. From the proof of Theorem 4.3, we see that the weight of the undetectable error considered is

$$\begin{aligned} w_q(\mathbf{x}|c(\mathbf{x})) - w_q(\mathbf{y}|c(\mathbf{y})) &= w_q(\mathbf{x}) - w_q(\mathbf{y}) + w_q(c_2(\mathbf{x})) - w_q(c_2(\mathbf{y})) \\ &= w_q(\mathbf{x}) - w_q(\mathbf{y}) + w_q(\langle [u]_\theta \rangle_\rho) - w_q(\langle [u-\lambda)]_\theta \rangle_\rho) \\ &= j\mu - \lambda + w_q(\langle \lambda \rangle_\rho) \\ &= j\mu - \lambda + w_q(\lambda). \end{aligned}$$

Suppose that

$$\lambda = \sum_{i=0}^{\rho-1} a_i q^i, \text{ and } \lambda' = \sum_{i=0}^{\rho-1} a'_i q^i,$$

where $a_i, a'_i \in F$ and $\langle \lambda \rangle_\rho \subseteq \langle \lambda' \rangle_\rho$, that is, $a_i \leq a'_i$ for all i. Then

$$(\lambda' - w_q(\lambda')) - (\lambda - w_q(\lambda)) = \sum_{i=0}^{\rho-1} (a'_i - a_i)(q^i - 1) \geq 0. \tag{4.9}$$

We note that $\langle \theta - 1 \rangle_\rho = (q-1, q-1, \ldots, q-1)$. Hence $w_q(\langle \theta - 1 \rangle_\rho) = (q-1)\rho$ and $\langle [u]_\theta \rangle_\rho \subseteq \langle \theta - 1 \rangle_\rho$. By (4.9), if $\langle \lambda \rangle_\rho \subseteq \langle [u]_\theta \rangle_\rho$, then

$$\lambda - w_q(\lambda) \leq [u]_\theta - w_q([u]_\theta) \leq \theta - 1 - (q-1)\rho. \tag{4.10}$$

Therefore, the weight of an uncorrectable error is lower bounded as follows:

$$j\mu - (\lambda - w_q(\lambda)) \geq \mu - \theta + 1 + (q-1)\rho.$$

Consider the uncorrectable errors of minimal weight. We see that we get equality in (4.9) if and only if $a'_i = a_i$ for all $i \geq 1$. Hence we have equality in both places in (4.10) if and only if

$$\lambda = \theta - \epsilon \text{ and } [u]_\theta = \theta - \delta,$$

where $0 \leq \delta \leq \epsilon \leq q-1$. This proves the following theorem

Theorem 4.4. *A GBL-code detects all errors of weight up to*

$$(\sigma - 1)\theta + (q-1)\rho,$$

and there are undetectable errors of weight

$$(\sigma - 1)\theta + (q-1)\rho + 1.$$

Undetectable errors of minimal weight occur exactly for code words of coweight $t\theta - \delta$ for $t \geq 1$ and $0 \leq \delta \leq q-1$. For such code words, an error is undetectable if the weight of the error to the information part is $\mu - \theta - \epsilon + \delta$ where $\delta \leq \epsilon \leq q-1$ and the last ρ symbols of the check part, namely $(q-1, q-1, \ldots, q-1, q-1-\delta)$, are changed to $(0, 0, \ldots, 0, q-1-\epsilon)$.

A natural question is: given q and r, which value of ω maximizes

$$A(q, r, \omega) = (\sigma - 1)\theta + (q-1)\rho.$$

For $q = 2$ it was shown by a simple proof in Kløve, Oprisan, and Bose (2005b) that the maximum is obtained for $\omega = 4$ when $r \geq 5$. For $q \geq 3$ it seems to be much more complicated to answer the question. Numerical computations reported in Gancheva and Kløve (2005b) indicate that for $q \geq 3$, the maximum is obtained for $\omega = 2$. The computations show that $A(q, r, 2) > A(q, r, \omega)$ for $3 \leq q \leq 7$ and $\omega \leq 100$. One can also show the following result.

Theorem 4.5. *Let $q \geq 3$. We have $A(q, r, 2) > A(q, r, 1)$ for $r \geq 3$. For $3 \leq \omega \leq 6$ we have $A(q, r, \omega - 1) > A(q, r, \omega)$ for $r \geq \omega$.*

This shows that for $\omega \leq 6$, the maximum is obtained for $\omega = 2$ and that for $\omega \geq 2$, the value of $A(q, r, \omega)$ decreases with ω. Whether this is true also for $\omega > 6$ is an open question.

4.1.3 *Diversity combining on the Z-channel*

For noisy asymmetric channels repeated retransmissions can increase the reliability, at the cost of decreasing the throughput efficiency of the system. One such method is called *diversity combining*. The scheme works as follows. Let a code word $\mathbf{x} = (x_1, x_2, \ldots, x_n)$ be transmitted repeatedly. Let the received word in the rth transmission be denoted by $\mathbf{x}_r = (x_{r,1}, x_{r,2}, \ldots, x_{r,n})$. At the receiving end we store a vector $\mathbf{z}$. We denote the value of $\mathbf{z}$ after the rth transmission by $\mathbf{z}_r = (z_{r,1}, z_{r,2}, \ldots, z_{r,n})$. It is computed by

$$\begin{aligned} z_{0,i} &= 0, \\ z_{r,i} &= \max(z_{r-1,i}, x_{r,i}) \end{aligned}$$

for $1 \le i \le n$. We note that

$$z_{r-1,i} \le z_{r,i} \le x_i.$$

When the combined word $\mathbf{z}$ becomes a code word, this is passed on, and a new code word is transmitted. If the passed-on code word is different from the one sent, then we have an undetected error. We will further assume that there is a limit k on the number of transmissions of a code word, that is, if the combined word is not a code word after k transmissions, it is discarded. A special case of the protocol is a protocol without a limit on the number of transmissions (that is, $k = \infty$).

The probability that the combined word is passed on with an undetected error will depend on the channel, the code word transmitted $\mathbf{x}$, the set $X = X_{\mathbf{x}}$ of code words $\mathbf{y}$ such that $\mathbf{y} \subset \mathbf{x}$ (that is, $y_i \le x_i$ for all i and $\mathbf{x} \ne \mathbf{y}$), and the maximum number of transmissions k. We will here discuss in some detail the situation when the channel is the Z-channel (binary asymmetric channel) with transition probability p, and we write $P_k(\mathbf{x}, X; p)$ for the probability of passing on a wrong code word (an undetected error).

We assume that $\mathbf{x} \ne \mathbf{0}$. Note that

$$\begin{aligned} &P_k(\mathbf{x}, X; 0) = 0, \\ &P_k(\mathbf{x}, X; 1) = 0, \text{ if } \mathbf{0} \in X, \\ &P_k(\mathbf{x}, X; 1) = 1, \text{ if } \mathbf{0} \notin X. \end{aligned}$$

Therefore, from now on we will assume that $0 < p < 1$. If we introduce more code words into X, P_k will increase, that is, if $X \subset Y$, then $P_k(\mathbf{x}, X; p) < P_k(\mathbf{x}, Y; p)$. One extreme case is when all $\mathbf{y} \subset \mathbf{x}$ are code words. Then $P = 1 - (1-p)^{w_H(\mathbf{x})}$. The other extreme is when X is empty: $P_k(\mathbf{x}, \emptyset; p) = 0$.

The case when X contains a single code word

We first consider the case when X contains exactly one code word, that is, there is exactly one code word $\mathbf{y}$ such that $\mathbf{y} \subset \mathbf{x}$. Let $w = w_H(\mathbf{x})$,

$u = w_H(\mathbf{y})$ and $d = w - u$. Since $P_k(\mathbf{x}, \{\mathbf{y}\}; p)$ only depends on w, u, p, and k, we write $P_k(w, u; p)$.

Suppose that the combined word becomes $\mathbf{y}$ after exactly r transmissions. The d positions not in the support of $\mathbf{y}$ must be in error for all r transmissions and the probability of this happening is p^{dr}. The u positions in the support of $\mathbf{y}$ must become all 1 for the first time after exactly r transmissions. Consider one position. The probability that the 1 in this position is transformed to a zero in all the r transmissions is p^r. Hence the probability that the bit in this position in $\mathbf{z}_r$ is a one is $1 - p^r$. Hence the probability that all the u positions in the support of $\mathbf{y}$ are one after r transmissions is $(1 - p^r)^u$. The probability that this happens for the first time after *exactly* r transmissions is therefore $(1-p^r)^u - (1-p^{r-1})^u$. Hence, the probability that the $\mathbf{z}_r = \mathbf{y}$ after exactly r transmissions is

$$p^{dr}[(1 - p^r)^u - (1 - p^{r-1})^u]. \tag{4.11}$$

Summing over all r we get

$$P_k(w, u; p) = \sum_{r=1}^{k} p^{dr}[(1 - p^r)^u - (1 - p^{r-1})^u].$$

In particular (if $k \geq d$),

$$P_k(w, u; p) = p^d(1 - p)^u + O(p^{2d}).$$

We can rewrite the expression for $P_k(w, u; p)$:

$$\begin{aligned} P_k(w, u; p) &= \sum_{r=1}^{k} p^{dr} \Big[\sum_{j=0}^{u} \binom{u}{j} (-1)^j p^{jr} - \sum_{j=0}^{u} \binom{u}{j} (-1)^j p^{j(r-1)} \Big] \\ &= \sum_{j=0}^{u} \binom{u}{j} (-1)^{j-1} (1 - p^j) p^d \sum_{r=1}^{k} p^{(d+j)(r-1)} \\ &= \sum_{j=0}^{u} \binom{u}{j} (-1)^{j-1} (1 - p^j) p^d \frac{1 - p^{k(d+j)}}{1 - p^{d+j}}. \end{aligned}$$

The values of $P_k(\mathbf{x}, X; p)$ when the code words in X are unordered

We say that the code words of X are unordered if $\mathbf{y} \not\subset \mathbf{y}'$ and for all $\mathbf{y}, \mathbf{y}' \in X$, $\mathbf{y} \neq \mathbf{y}' \in X$. We observe that the event that the combined word becomes $\mathbf{x}$ after having been $\mathbf{y}$ and the event that it becomes $\mathbf{x}$ after having been $\mathbf{y}'$, where $\mathbf{y} \neq \mathbf{y}'$, are mutually exclusive. Hence

$$P_k(\mathbf{x}, X; p) = \sum_{\mathbf{y} \in X} P_k(\mathbf{x}, \{\mathbf{y}\}; p).$$

In the special case where all the code words in X have the same weight, u, (and $\mathbf{x}$ has weight w) we get

$$P_k(\mathbf{x}, X; p) = |X| \sum_{j=0}^{u} \binom{u}{j} (-1)^{j-1}(1-p^j)p^d \frac{1-p^{k(d+j)}}{1-p^{d+j}}.$$

Bounds on $P_k(\mathbf{x}, X; p)$ when some code words in X cover others

In the general case when the code words in X are not unordered, to determine $P_k(\mathbf{x}, X; p)$ becomes more complex and may even be unfeasible. Therefore, it is useful to get some bounds. Let T be the smallest value of $w_H(\mathbf{x}) - w_H(\mathbf{v})$ over all code words $\mathbf{v}$ and $\mathbf{y}$ such that $\mathbf{v} \subset \mathbf{y} \subset \mathbf{x}$, (in particular, $T \geq 2d$ where d is the minimum distance of the code). Define the set Y by

$$Y = \{\mathbf{y} \in X \mid w_H(\mathbf{y}) > w_H(\mathbf{x}) - T\}.$$

Then the code words of Y are independent. Hence $P_k(\mathbf{x}, Y; p)$ can be computed as explained above. Further, if p^T is small, then $P_k(\mathbf{x}, Y; p)$ is a good approximation for $P_k(\mathbf{x}, X; p)$. We will make this last claim more precise. On the one hand we know that $P_k(\mathbf{x}, X; p) \geq P_k(\mathbf{x}, Y; p)$. On the other hand, if the combined word becomes a code word not in Y, then after the first transmission, the combined word must have weight $w - T$ or less. The probability that the combined word is *some* vector of weight $w - T$ or less (not necessary a code word) after the first transmission is

$$\sum_{j=0}^{w-T} \binom{w}{j} p^{w-j}(1-p)^j \leq p^T(1-p)^{w-T} \sum_{j=0}^{w-T} \binom{w}{j} < p^T(1-p)^{w-T}2^w.$$

Therefore

$$P_k(\mathbf{x}, Y; p) \leq P_k(\mathbf{x}, X; p) < P_k(\mathbf{x}, Y; p) + 2^w p^T(1-p)^{w-T}.$$

For the whole code C we get

$$P_k(C; p) = \frac{1}{|C|} \sum_{\mathbf{x} \in C} P_k(\mathbf{x}, X_{\mathbf{x}}; p).$$

4.2 Coding for a symmetric channel with unknown characteristic

If we have to transmit over a symmetric channel with unknown characteristics, the best strategy is to do a coding which makes the number of undetectable errors as small as possible for all possible error patterns. To be more precise, let us first consider a channel transmitting binary symbols. Let C be an $(n, M; 2)$ code. If $\mathbf{x} \in C$ is transmitted and $\mathbf{y} \in C$ is received, then $\mathbf{e} = \mathbf{y} + \mathbf{x}$ is the error pattern. The error is undetected if (and only if) $\mathbf{y} = \mathbf{x} + \mathbf{e} \in C$.

For each error pattern $\mathbf{e} \in F_2^n$, let

$$Q(\mathbf{e}) = \#\{\mathbf{x} \in C \mid \mathbf{x} + \mathbf{e} \in C\}, \tag{4.12}$$

that is, $Q(\mathbf{e})$ is the number of code words in C for which $\mathbf{e}$ will be an undetectable error. The idea is to choose C such that

$$Q(C) = \max\{Q(\mathbf{e}) \mid \mathbf{e} \in F_2^n \setminus \{\mathbf{0}\}\}$$

is as small as possible.

For channels with q elements, where q is a prime power, we consider the same approach. We assume that $F_q = GF(q)$ and when $\mathbf{x} \in GF(q)^n$ is submitted and we have an error pattern $\mathbf{e} \in GF(q)^n$, then $\mathbf{x} + \mathbf{e} \in GF(q)^n$ is received.

4.2.1 *Bounds*

For a code $C \subseteq GF(q)^n$ and an error pattern $\mathbf{e} \in GF(q)^n$, let

$$Q(\mathbf{e}) = \#\{\mathbf{x} \in C \mid \mathbf{x} + \mathbf{e} \in C\}, \tag{4.13}$$

and

$$Q(C) = \max\{Q(\mathbf{e}) \mid \mathbf{e} \in GF(q)^n \setminus \mathbf{0}\}. \tag{4.14}$$

Theorem 4.6. *If C is an $(n, M; q)$ code and $M \geq 2$, then $Q(C) \geq 1$.*

Proof. By assumption, C has two distinct code words, $\mathbf{x}$ and $\mathbf{y}$, say. Let $\mathbf{e} = \mathbf{y} - \mathbf{x}$. Then $Q(\mathbf{e}) \geq 1$. Hence $Q(C) \geq 1$. □

Theorem 4.7. *If C is an $(n, M; q)$ code, where q is even, and $M \geq 2$, then $Q(C) \geq 2$.*

Proof. By assumption, C has two distinct code words, $\mathbf{x}$ and $\mathbf{y}$, say. Let $\mathbf{e} = \mathbf{y} - \mathbf{x}$. Then $\mathbf{x} - \mathbf{y} = \mathbf{e}$. Hence $Q(\mathbf{e}) \geq 2$ and so $Q(C) \geq 2$. □

Theorem 4.8. *If C is an $(n, M; q)$ code, then $M(M-1) \leq (q^n - 1)Q(C)$.*

Proof. The number of pairs $(\mathbf{x}, \mathbf{y})$ of distinct elements in C is $M(M-1)$. For each of the possible $q^n - 1$ error patterns $\mathbf{e} \in GF(q)^n \setminus \{\mathbf{0}\}$, there are at most $Q(C)$ pairs $(\mathbf{x}, \mathbf{y}) \in C^2$ such that $\mathbf{y} - \mathbf{x} = \mathbf{e}$. Hence $M(M-1) \leq (q^n - 1)Q(C)$. □

We will first consider systematic codes C for which $Q(C) = 1$ and q is odd. Since

$$q^{2k-1} < q^k(q^k - 1) < q^{2k} - 1$$

we get the following corollary.

Corollary 4.1. *If C is a systematic $(n, q^k; q)$ code and $Q(C) = 1$, then $n \geq 2k$.*

4.2.2 *Constructions*

Construction for q odd

Let q be an odd prime power. There is a natural linear bijection F from $GF(q)^k$ to $GF(q^k)$: let $\alpha \in GF(q^k)$ be a primitive root (that is, $\alpha^i \neq 1$ for $1 \leq i \leq q^k - 2$ and $\alpha^{q^k-1} = 1$) and define F by

$$F : (x_0, x_1, \ldots, x_{k-1}) \mapsto x = x_0 + x_1\alpha + x_2\alpha^2 + \cdots + \alpha^{k-1}x_{k-1}.$$

Note that $F(x_0, x_1, \ldots, x_{k-1}) = 0$ if and only if $(x_0, x_1, \ldots, x_{k-1}) = (0, 0, \ldots, 0)$. Define the $(2k, q^k; q)$ code by

$$C = \{(x_0, x_1, \ldots, x_{k-1}, y_0, y_1, \ldots, y_{k-1}) \mid (x_0, x_1, \ldots, x_{k-1}) \in GF(q)^k\}$$

where $(y_0, y_1, \ldots, y_{k-1})$ is defined by

$$F(y_0, y_1, \ldots, y_{k-1}) = F(x_0, x_1, \ldots, x_{k-1})^2.$$

Mapping C into $GF(q^k)^2$ by F, let the image be $\hat{C}$, that is

$$\hat{C} = \{(x, x^2) \mid x \in GF(q^k)\}.$$

This is clearly a systematic $(2, q^k; q^k)$ code over $GF(q^k)$. An error pattern $(e_0, e_1, \ldots, e_{k-1}, f_0, f_1, \ldots, f_{k-1})$ maps into $(e, f) \in GF(q^k)^2$, where

$$e = F(e_0, e_1, \ldots, e_{k-1}) \text{ and } f = F(f_0, f_1, \ldots, f_{k-1}),$$

and

$$(e_0, e_1, \ldots, e_{k-1}, f_0, f_1, \ldots, f_{k-1}) \neq (0, 0, \ldots, 0)$$

if and only if $(e, f) \neq (0, 0)$.

Let $(e, f) \neq (0, 0)$ be an error pattern and suppose $(x, x^2) + (e, f) \in \hat{C}$, that is

$$(x + e)^2 = x^2 + f.$$

Since $(x + e)^2 = x^2 + 2xe + e^2$, this implies that

$$2xe + e^2 = f. \tag{4.15}$$

If $e = 0$, (4.15) implies that $f = 0$, that is $(e, f) = (0, 0)$; however this is not the case for an error pattern. Hence $e \neq 0$ and so $x = (2e)^{-1}(f - e^2)$, that is, x is uniquely determined and so $Q((e, f)) = 1$. Hence $Q(C) = Q(\hat{C}) = 1$.

Construction for q even

Construction 1 can be modified to give a systematic $(2k, q^k; q)$ code C with $Q(C) = 2$. We give the details only for those parts at the construction and proof that differs.

Let F be the linear bijection from $GF(q)^k$ to $GF(q^k)$ defined in Construction 1. Define the $(2k, q^k; q)$ code C by

$$\hat{C} = \{(x, x^3) \mid x \in GF(q^k)\}.$$

As for Construction 1, an error pattern $(e_0, e_1, \ldots, e_{k-1}, f_0, f_1, \ldots, f_{k-1})$ maps into (e, f) where

$$(e_0, e_1, \ldots, e_{k-1}, f_0, f_1, \ldots, f_{k-1}) \neq (0, 0, \ldots, 0)$$

if and only if $(e, f) \neq (0, 0)$.

Let $(e, f) \neq (0, 0)$ be an error pattern and suppose $(x, x^3) + (e, f) \in \hat{C}$, that is

$$(x + e)^3 = x^3 + f.$$

Since $(x + e)^3 = x^3 + x^2e + xe^2 + e^3$, this implies that

$$x^2e + xe^2 + e^3 = f. \tag{4.16}$$

If $e = 0$, (4.16) implies that $f = 0$, that is $(e, f) = (0, 0)$; however this is not the case for an error pattern. Hence $e \neq 0$. The equation (4.16) therefore has at most two solutions for x. Hence $Q((e, f)) \leq 2$ and $Q(C) = Q(\hat{C}) = 2$.

4.3 Codes for detection of substitution errors and transpositions

A *transposition* occurs when two neighbouring elements change places. For example, *acb* is obtained from *abc* by transposition of the two last elements. Transpositions are a common type of error in data that is handled manually. Therefore, there has been introduced many codes to detect substitution errors and/or transpositions, in particular for digital (base 10) data.

4.3.1 *ST codes*

An ST (substitution - transposition) code is a code that can detect a single substitution error or transposition error. We first give a simple upper bound on the size of ST codes.

Theorem 4.9. *If C is an $(n, M; q)$ ST code, then $M \leq q^{n-1}$.*

Proof. If two vectors of length n are identical, except in the last position, then one is obtained from the other by a single substitution. Hence at most one of them can belong to C. Therefore, the code words are determined by the first $n-1$ elements and therefore the code contains at most q^{n-1} code words. □

It turns out that for all $q > 2$ and all n there exist $(n, q^{n-1}; q)$ ST codes. We describe some constructions of ST codes for various types of q.

Construction of codes over $GF(q)$ where $q > 2$

If q is a prime power, we can consider an $[n, n-1; q]$ code C. Let the dual code be generated by $(w_1, w_2, \ldots, w_n)$, that is

$$C = \Big\{(x_1, x_2, \ldots x_n) \in GF(q)^n \mid \sum_{i=1}^{n-1} w_i x_i = 0\Big\}. \tag{4.17}$$

The code can detect single substitution errors if all

$$w_i \neq 0 \tag{4.18}$$

(that is, $d(C) \geq 2$). We see that if x_i is changed to x'_i, then $\sum_{i=1}^{n-1} w_i x_i$ is changed by

$$-w_i x_i + w_i x'_i = w_i(x_i - x'_i) \neq 0.$$

Similarly, if x_i and x_{i+1} are transposed, then the sum is changed by

$$-w_i x_i - w_{i+1} x_{i+1} + w_i x_{i+1} + w_{i+1} x_i = (w_i - w_{i+1})(x_{i+1} - x_i) \neq 0$$

if $x_i \neq x_{i+1}$ and

$$w_i \neq w_{i+1}. \tag{4.19}$$

If $q > 2$, we can find w_i that satisfy conditions (4.18) and (4.19), for example $w_i = 1$ for even i and $w_i = a$ for odd i, where $a \notin \{0, 1\}$.

Construction of codes over Z_q where q is odd

If q is an odd integer, we can make a similar construction with elements from Z_p. We define

$$C = \Big\{(x_1, x_2, \ldots x_n) \mid \sum_{i=1}^{n} w_i x_i \equiv 0 \pmod q\Big\}$$

where now

$$\gcd(w_i, q) = 1 \text{ and } \gcd(w_{i+1} - w_i, q) = 1$$

for all i. The proof that this gives an ST code is similar. A possible choice for the w_i is $w_i = 1$ for even i and $w_i = 2$ for odd i.

Product construction

Let C_1 be an $(n, q_1^{n-1}; q_1)$ ST code over F_{q_1} and C_2 an $(n, q_2^{n-1}; q_2)$ ST code over F_{q_2}. Let $q = q_1 q_2$ and

$$F_q = F_{q_1} \times F_{q_2} = \{(a, b) \mid a \in F_{q_1} \text{ and } b \in F_{q_2}\}.$$

Define

$$C = \{((a_1, b_1), (a_2, b_2), \ldots, (a_n, b_n)) \mid \mathbf{a} \in C_1, \mathbf{b} \in C_2\}.$$

Then C is an ST code over F_q. To check this, we first note that if $(a_i', b_i') \neq (a_i, b_i)$, then $a_i' \neq a_i$ or $b_i' \neq b_i$ (or both). Hence a single substitution will change a code word to a non-code word. Similarly, if $(a_i, b_i) \neq (a_{i+1}, b_{i+1})$, then $a_i \neq a_{i+1}$ or $b_i \neq b_{i+1}$ or both. Hence, a transposition will change a code word to a non-code word.

Any integer q can be written as $2^m r$ where r is odd. If $m \geq 2$, Construction 1 gives an $(n, 2^{mn-1}; 2^m)$ ST code and Construction 2 gives an $(n, r^{n-1}; r)$ST code. Combining the two, using Construction 3, we get an $(n, q^{n-1}; q)$ code. It remains to consider q of the form $2r$, where r is odd. This case is more complicated.

Construction of codes over Z_q for $q \equiv 2 \pmod 4$

Let $q = 2r$, where $r > 1$ is odd. Let the alphabet be Z_q. For $\mathbf{x} = (x_1, x_2, \ldots, x_n) \in Z_q^n$, we define some auxiliary functions:

$n_e = n_e(\mathbf{x})$ is the number of even elements in $\mathbf{x}$,
$n_o = n_o(\mathbf{x})$ is the number of odd elements in $\mathbf{x}$,
$S_e(\mathbf{x}) = \sum_{i=1}^{n_e} (-1)^i e_i$ where $e_1, e_2, \ldots, e_{n_e}$ are the even elements of $\mathbf{x}$,
$S_o(\mathbf{x}) = \sum_{i=1}^{n_o} (-1)^i o_i$ where $o_1, o_2, \ldots, o_{n_o}$ are the odd elements of $\mathbf{x}$,
$K_n(x_1, x_2, \ldots, x_j)$ are the number of $i \leq j$
 such that x_i is even and $n - i$ is even,
$S(\mathbf{x}) = (-1)^n S_e(\mathbf{x}) + S_o(\mathbf{x}) + 2K_n(\mathbf{x})$.

In particular, we see that

$$K_n(x_1, x_2, \ldots, x_{n-1}) = K_n(x_1, x_2, \ldots, x_n). \tag{4.20}$$

Let

$$C = \{\mathbf{x} \in Z_q^n \mid S(\mathbf{x}) \equiv 0 \pmod q\}.$$

We will show that C is an ST code of size q^{n-1}. We break the proof up into a couple of lemmas.

Lemma 4.1. *For $\mathbf{c} = (c_1, c_2, \ldots c_{n-1})$, define c_n by*

$$-c_n \equiv (-1)^n S_e(\mathbf{c}) + S_o(\mathbf{c}) + 2K_n(\mathbf{c}) \pmod{q}.$$

Then

$$(\mathbf{c}|c_n) \in C.$$

Proof. Let $\mathbf{x} = (\mathbf{c}|c_n)$. First we note that $S_e(\mathbf{c})$ is even and $S_o(\mathbf{c}) \equiv n_o(\mathbf{c})$ (mod 2). Hence $c_n \equiv n_o(\mathbf{c}) \pmod{2}$ and so $n_o(\mathbf{x})$ is even. This also implies that $n_e(\mathbf{x}) \equiv n \pmod{2}$. If $n_o(\mathbf{c})$ is odd, then we get

$$S_o(\mathbf{x}) = S_o(\mathbf{c}) + (-1)^{n_o(\mathbf{x})} c_n = -(-1)^n S_e(\mathbf{c}) - 2K_n(\mathbf{c}),$$
$$S_e(\mathbf{x}) = S_e(\mathbf{c}).$$

Combining these and (4.20), we see that $\mathbf{x} \in C$.

If $n_o(\mathbf{c})$ is even, we similarly get

$$S_e(\mathbf{x}) = S_e(\mathbf{c}) + (-1)^{n_e(\mathbf{x})} c_n = -S_o(\mathbf{c}) - 2K_n(\mathbf{c}),$$
$$S_o(\mathbf{x}) = S_o(\mathbf{c}),$$

and we can conclude that $\mathbf{x} \in C$ also in this case. □

Lemma 4.2. *The code*

$$C = \{\mathbf{x} \in Z_q^n \mid S(\mathbf{x}) \equiv 0 \pmod{q}\}$$

is an ST code.

Proof. First we note that

$$n_o(\mathbf{x}) \equiv S_o(\mathbf{x}) \equiv 0 \pmod{2}$$

for all code words. We consider the various types of single errors that can occur.

- If a substitution error change the parity of a symbol, then $n_o(\mathbf{c})$ is changed by one and becomes odd.
- If a substitution error change an even element, say e_j to another even element e'_j, then $S_e(\mathbf{x})$ is changed to $S_e(\mathbf{x}) - (-1)^j(e_j - e'_j)$ whereas $S_o(\mathbf{x})$ and $K_n(\mathbf{x})$ are unchanged. Hence $S(\mathbf{x})$ is changed by $(-1)^{j+1}(e_j - e'_j) \not\equiv 0$ (mod q).
- If a substitution error change an odd element to another odd element, the situation is similar.
- If two elements of opposite partity are transposed, then $K_n(\mathbf{x})$ is changed by one, whereas $S_e(\mathbf{x})$ and $S_o(\mathbf{x})$ are unchanged. Hence $S(\mathbf{x})$ is changed by $\pm 1 \not\equiv 0 \pmod{q}$.

- If two even elements, e_j and e_{j+1} are transposed, then $S_e(\mathbf{x})$ is changed by

$$-(-1)^j(e_j - e_{j+1}) + (-1)^j(e_{j+1} - e_j) = 2(-1)^j(e_{j+1} - e_j)$$

  whereas $S_o(\mathbf{x})$ and $K_n(\mathbf{x})$ are unchanged. Hence $S(\mathbf{x})$ is changed by $2(-1)^j(e_{j+1} - e_j)$. Since $0 < |e_{j+1} - e_j| < q = 2r$ and $e_{j+1} - e_j$ is even, we can conclude that $e_{j+1} - e_j \not\equiv 0 \pmod{r}$ and so $2(-1)^j(e_{j+1} - e_j) \not\equiv 0 \pmod{q}$.
- If two odd elements are transposed, the situation is similar.

□

Since C is an ST code, its size is at most q^{n-q} by Theorem 4.9. By Lemma 4.1, the size is exactly q^{n-1}.

Verhoeff's code

Another code construction for the case when $q = 2r$, $r > 1$ and odd, is based on so-called dihedral groups D_r. This was first shown by Verhoeff for $q = 10$. We describe this code, but omit the proof that it is an ST code.

One can map the elements of D_5 onto Z_{10}. The "addition" table of $i \oplus j$ is given in Table 4.1. An entry in the table gives $i \oplus j$ where i is given in the left column and j in the top row. In particular, $i \oplus 0 = 0 \oplus i = i$ for all i, that is, 0 is the unit element of the group. The operation $\oplus$ is not commutative. For example $1 \oplus 5 = 6$ and $5 \oplus 1 = 9$. We see that if $i \oplus j = 0$, then $j \oplus i = 0$, that is, j is the inverse of i in the group.

Table 4.1 Addition table for $\oplus$.

0	1	2	3	4	5	6	7	8	9
1	2	3	4	0	6	7	8	9	5
2	3	4	0	1	7	8	9	5	6
3	4	0	1	2	8	9	5	6	7
4	0	1	2	3	9	5	6	7	8
5	9	8	7	6	0	4	3	2	1
6	5	9	8	7	1	0	4	3	2
7	6	5	9	8	2	1	0	4	3
8	7	6	5	9	3	2	1	0	4
9	8	7	6	5	4	3	2	1	0

For the construction, we also do a substitution in each position using Table 4.2 of functions $f_i : Z_{10} \to Z_{10}$ where i are given in the left column and x in the top row. For $i \geq 8$, $f_i = f_{i-8}$. Hence, for example, $f_{26} = f_2$. The code is defined by

Table 4.2 Functions f_i for Verhoeff's construction.

$f_i(x)$	0	1	2	3	4	5	6	7	8	9
0	0	1	2	3	4	5	6	7	8	9
1	1	5	7	6	2	8	3	0	9	4
2	5	8	0	3	7	9	6	1	4	2
3	8	9	1	6	0	4	3	5	2	7
4	9	4	5	3	1	2	6	8	7	0
5	4	2	8	6	5	7	3	9	0	1
6	2	7	9	3	8	0	6	4	1	5
7	7	0	4	6	9	1	3	2	5	8

$$\{(x_1, x_2, \ldots, x_n) \mid f_1(x_1) \oplus f_2(x_2) \oplus \cdots \oplus f_n(x_n) = 0\}.$$

Analysis has shown that this code detects all single substitutions and all single transpositions. Also more than 95% of the twin errors are detected.

Construction of a binary code

It turns out that the maximal size of a binary ST code of length n is $\left\lceil \frac{2^n}{3} \right\rceil$. We will not give the proof of this fact here, but give a code of this size. By Construction 2, the ternary code

$$\left\{(x_1, x_2, \ldots, x_n) \in Z_3^n \mid \sum_{i=1}^{n}(-1)^i x_i \equiv 0 \pmod 3\right\}$$

is an ST code. A binary subcode is the code

$$\left\{(x_1, x_2, \ldots, x_n) \in Z_2^n \mid \sum_{i=1}^{n}(-1)^i x_i \equiv 0 \pmod 3\right\}.$$

As a subcode, this is also an ST code, and it can be shown that it has maximal size, that is $\left\lceil \frac{2^n}{3} \right\rceil$.

4.3.2 *ISBN*

A number of modifications of the constructions given in the previous section are used in practical systems. A typical example is the *International Standard Book Number* - ISBN. The traditional ISBN code is a ten-digit number. The first nine digits $x_1 x_2 \ldots x_9$ codes information about the country, publisher and the individual book. The last digit x_{10} is a check digit determined with check vector $(10, 9, \ldots, 1)$ modulo 11. This code can detect any single substitution or transposition error. Note, however, that

$w_5 + w_6 = 6 + 5 \equiv 0 \pmod{11}$. Hence twin errors in positions 5 and 6 are never detected. A (digital) information vector $(x_1, x_2, \dots, x_9) \in Z_{10}^9$ determines a check symbol $x_{10} \in Z_{11}$. If $x_{10} = 10$, the check symbol is written X. Some countries have chosen to avoid the symbol X by not using the information vectors that have X as check symbol.

From January 1, 2007, a new ISBN code with 13 digits has been introduced. The first three digits $y_1y_2y_3$ are 978 (later 979 will also be used), the next nine digits $y_4y_5\dots y_{12}$ are now the information, and the last digit y_{13} is the check digit. The code is determined by the check vector $(1,3,1,3,\dots,1,3,1)$ modulo 10. This choice of check digit is the same as for the bar code EAN described below. In fact, most books published in recent years contains both the ten digit ISBN number and a bar code with the new 13 digit ISBN. If you look at one, you will notice that the check digits are (usually) not the same. Since $\gcd(1,10) = \gcd(3,10) = 1$, this code can detect all single substitution errors. However, transposition of x_j and x_{j+1} will not be detected if $x_{j+1} \equiv x_j + 5 \pmod{10}$.

4.3.3 *IBM code*

The IBM code is also known as the Luhn code after its inventor. It is usually presented as a code over Z_{10}, but here we consider the more general alphabet of size $q = 2r$ where $r > 1$ is odd. We gave two constructions for this alphabet size in the previous section. The IBM code is simpler, but it cannot detect all transpositions.

For a sequence $(x_1, x_2, \dots, x_{n-1})$ a check symbol x_n is chosen such that

$$\sum_{i=1}^{n} y_i \equiv 0 \pmod{q}$$

where

$$\begin{aligned} y_i &= x_i && \text{for } i = n, n-2, n-4, \dots, \\ y_i &= 2x_i && \text{for } i = n-1, n-3, n-5, \dots \text{ and } 0 \le x_i \le r-1, \\ y_i &= 2x_i + 1 && \text{for } i = n-1, n-3, n-5, \dots \text{ and } r \le x_i \ge q-1. \end{aligned}$$

Note that y_i runs through Z_q when x_i does. Hence, any single substitution error can be detected. Consider a transposition of x_i and x_{i+1} where $n - i$ is even. Then the sum $\sum_{i=1}^{n} y_i$ is changed by

$$\begin{aligned} -(x_i + 2x_{i+1}) + (x_{i+1} + 2x_i) &= x_i - x_{i+1} && \text{if } x_i < r \text{ and } x_{i+1} < r, \\ -(x_i + 2x_{i+1} + 1) + (x_{i+1} + 2x_i) &= x_i - x_{i+1} - 1 && \text{if } x_i < r \text{ and } x_{i+1} \ge r, \\ -(x_i + 2x_{i+1}) + (x_{i+1} + 2x_i + 1) &= x_i - x_{i+1} + 1 && \text{if } x_i \ge r \text{ and } x_{i+1} < r, \\ -(x_i + 2x_{i+1} + 1) + (x_{i+1} + 2x_i + 1) &= x_i - x_{i+1} && \text{if } x_i \ge r \text{ and } x_{i+1} \ge r. \end{aligned}$$

We see that the change is not congruent zero modulo q, except in two cases, namely $(x_i, x_{i+1}) = (0, q-1)$ and $(x_i, x_{i+1}) = (q-1, 0)$. Hence, the code detects all transpositions, except transpositions of 0 and $q-1$.

4.3.4 *Digital codes with two check digits*

In some applications there are other types of errors in addition to ST errors that are relatively frequent and therefore we want codes to detect such errors.

We give one example modeled the Norwegian personal number codes. In Norway each person is assigned a unique 11 digit number $p_1p_2p_3p_4p_5p_6p_7p_8p_9p_{10}p_{11}$ where $p_1p_2p_3p_4p_5p_6$ is the date of birth in the form *ddmmyy*, $p_7p_8p_9$ is the persons serial number and $p_{10}p_{11}$ are check digits. In addition to simple substitutions, twin errors, and transpositions, common types of errors are interchanging date and month (i.e. $p_1p_2 \leftrightarrow p_3p_4$) and interchanging date and year ($p_1p_2 \leftrightarrow p_5p_6$, that is, $ddmmyy \leftrightarrow yymmdd$). Also, interchanging month and year happens. The check digits are chosen using a weighted sum modulo 11, discarding serial numbers which produce 10 as value for one or both of the check digits. How may the weights be chosen? If we choose the weights

$$\begin{pmatrix} 10 & 9 & 8 & 7 & 6 & 5 & 4 & 3 & 2 & 1 & 0 \\ 1 & 1 & 1 & 1 & 1 & 1 & 1 & 1 & 1 & 1 & 1 \end{pmatrix} \tag{4.21}$$

we know that all single substitutions, twin errors, and transpositions are detected. However, if $p_1 + p_2 \equiv p_3 + p_4 \pmod{11}$ interchanging date and month will not be detected. We modify (4.21) as follows: Suppose that we multiply the first column of (4.21) by s and the third column by t to get

$$\begin{pmatrix} 10s & 9 & 8t & 7 & 6 & 5 & 4 & 3 & 2 & 1 & 0 \\ s & 1 & t & 1 & 1 & 1 & 1 & 1 & 1 & 1 & 1 \end{pmatrix}$$

As long as both s and t are non-zero modulo 11, we can still detect all single substitutions, twin errors, and transpositions (from one or the other of the check digits). When p_1p_2 and p_3p_4 are interchanged, the sum defining the first check digit is changed by

$$\begin{aligned} &-(10sp_1 + 9p_2 + 8tp_3 + 7p_4) + (10sp_3 + 9p_4 + 8tp_1 + 7p_2) \\ &= (10s - 8t)(p_3 - p_1) + 2(p_4 - p_2). \end{aligned} \tag{4.22}$$

Similarly, the sum defining the second check digit is changed by

$$= (s - t)(p_3 - p_1). \tag{4.23}$$

We want at least one of these to be non-zero (modulo 11) if $p_1p_2 \neq p_3p_4$. This is guaranteed to be the case if $s \not\equiv t \pmod{11}$. In this case, the sum defining the last check digit is non-zero unless $p_1 = p_3$. On the other hand, if $p_1 = p_3$, then $p_2 \neq p_4$ and the change in the first sum is non-zero.

Similarly, if $t \not\equiv 1 \pmod{11}$ we are guaranteed to detect any transposition of month and year, and if $s \not\equiv 1 \pmod{11}$ any transposition of date and year. For example, $s = 3$ and $t = 2$ gives the check matrix

$$\begin{pmatrix} 8\,9\,5\,7\,6\,5\,4\,3\,2\,1\,0 \\ 3\,1\,2\,1\,1\,1\,1\,1\,1\,1\,1 \end{pmatrix}$$

of a code which can detect all single substitution errors, twin errors, transpositions, or the interchange of any two of the date, month, or year.

The weights actually used in the Norwegian system are different, namely

$$\begin{pmatrix} 3\,7\,6\,1\,8\,9\,4\,5\,2\,1\,0 \\ 5\,4\,3\,2\,7\,6\,5\,4\,3\,2\,1 \end{pmatrix}$$

4.3.5 *Barcodes*

There are a number of variants of barcodes. They usually have error detection on two levels, both in the coding of the number itself with a check digit and in the coding of the individual digits as a sequence of bars. As an illustration, we consider the important *European Article Numbering* EAN-13. The check digits are the same as for the new ISBN, that is, the weights are $(1,3,1,3,1,3,1,3,1,3,1,3,1)$ modulo 10. We denote the 13 digits number by $(x_1, x_2, x_3, \ldots, x_{13})$.

Table 4.3 Symbol encoding for barcodes

	left A	left B	right
0	0001101	0100111	1110010
1	0011001	0110011	1100110
2	0010011	0011011	1101100
3	0111101	0100001	1000010
4	0100011	0011101	1011100
5	0110001	0111001	1001110
6	0101111	0000101	1010000
7	0111011	0010001	1000100
8	0110111	0001001	1001000
9	0001011	0010111	1110100

Each digit, except the first x_1, are encoded into seven bars (black or white) of the same width. In this presentation, we represent the bars by 1 (black) and 0 (white). The total barcode starts with lead sequence 101, then barcodes for the six digits $x_2, x_3, x_4, x_5, x_6, x_7$ in the left part (that is 42 bars) follows, then a separator 01010, then the barcodes for the six digits $x_8, x_9, x_{10}, x_{11}, x_{12}, x_{13}$ in the right part, and finally a trailer sequence 101. The bars for the lead, the separator, and the trailer are longer than the others.

The seven bars encoding a digit contains two runs of zeros and two runs of ones. The codes for the digits in the left part all starts with 0 and ends with 1. For the digits in the right part, the codes starts with 1 and ends with 0. The encoding is done using Table 4.3.

For the six digits in the right part, the codes are listed under the heading "right". For the six digits in the left part, the encoding is done either with the code listed under "left A" or the one listed under "left B". The choice between the two is determined by x_1, using Table 4.4.

Table 4.4 Choice of A or B.

x_1	0	1	2	3	4	5	6	7	8	9
x_2	A	A	A	A	A	A	A	A	A	A
x_3	A	A	A	A	B	B	B	B	B	B
x_4	A	B	B	B	A	B	B	A	A	B
x_5	A	A	B	B	A	A	B	B	B	A
x_6	A	B	A	B	B	A	A	A	B	B
x_7	A	B	B	A	B	B	A	B	A	A

We illustrate with one example.

Example 4.3. Suppose that

$$x_1 x_2 \ldots x_{12} = 978981270586.$$

The check digit x_{13} is determined by

$$x_{13} \equiv -\{9+3\cdot 7+8+3\cdot 9+8+3\cdot 1+2+3\cdot 7+0+3\cdot 5+8+3\cdot 6\} = -140 \equiv 0 \pmod{10}.$$

The barcode for 9789812705860 is given in Table 4.5 over two lines and where the encoding is marked with A, B, and R (right).

Table 4.5 Barcode example.

	7 (A)	8 (B)	9 (B)	8 (A)	1 (B)	2 (A)	
101	0111011	0001001	0010111	0110111	0110011	0010011	01010
	7 (R)	0 (R)	5 (R)	8 (R)	6 (R)	0 (R)	
	1000100	1110010	1001110	1001000	1010000	1110010	101

4.4 Error detection for runlength-limited codes

For this section we find it convenient to introduce a different notation for binary sequences. First, 0^a denotes a sequence of a zeros. Any binary sequence of weight w, say, can then be represented as

$$0^{b_0}10^{b_1}10^{b_1}1\cdots 0^{b_{w-1}}10^{b_w}. \tag{4.24}$$

The binary sequence (4.24) is said to be a (d,k) constrained sequence if

$$b_0 \geq 0,\ b_w \geq 0, \quad \text{and } d \leq b_i \leq k \text{ for } 1 \leq i \leq w-1$$

where $0 \leq d \leq k \leq \infty$ ($k = \infty$ means that b_i may be arbitrarily large). For example,

$$0^2 10^2 10^4 10^7 10^3 1 = 001001000010000000010001$$

is a $(2,7)$ constrained sequence. Let $R_{n,d,k}$ denote the set of all (d,k) constrained sequences of length n. A (d,k) run-length limited code of length n is some subset of $R_{n,d,k}$. Both the set of even weight code words in $R_{n,d,k}$ and the set of odd weight code words in $R_{n,d,k}$ are codes that clearly can detect single bit errors, and the larger of the two has size at least $\frac{1}{2}\#R_{n,d,k}$.

Run-length limited codes have their main application in magnetic recording. One type of errors occurring are substitution errors. Another type of errors which is common in this context are *shifts*, that is, a one is moved one place to the left or to the right. In the terminology of the previous section, a shift is the same as a transposition.

Construction of a non-systematic code

The set $R_{n,d,k}$ can be partitioned into three subcodes that can detect any single bit-error or shift. Hence there exists such a code of size at least $\frac{1}{3}\#R_{n,d,k}$. The partition is as follows.

For a sequence $\mathbf{c}$ represented as (4.24), let β_i be the parity of b_i, that is $\beta_i \equiv b_i \pmod 2$ and $\beta_i \in \{0,1\}$. Further, define a sequence

$$s_{-1} = 0, s_0, s_1, \ldots s_w$$

of numbers from $\{0, 1, 2\}$ by

$$s_i \equiv ((2 - \beta_i)s_{i-1} + \beta_i) \pmod 3 \text{ for } 0 \le i \le w. \tag{4.25}$$

This is called the *check sequence* of $\mathbf{c}$. Note that $2 - \beta_i \not\equiv 0 \pmod 3$. Hence, if $s'_{i-1} \not\equiv s_{i-1} \pmod 3$, and $s'_i \equiv ((2 - \beta_i)s_{i-1} + \beta_i) \pmod 3$, then $s'_i \not\equiv s_i \pmod 3$. The *error check value* of the sequence is defined by $s(\mathbf{c}) = s_w$. The couple of examples in Table 4.6 illustrate this. We see that $s(\mathbf{c}_1) = 1$ and $s(\mathbf{c}_2) = 0$.

Table 4.6 Examples of check values.

$\mathbf{c}_1 = 0^2 1010^3 10^2$	$b_0 = 2,$	$b_1 = 1,$	$b_2 = 3,$	$b_3 = 2,$	
	$\beta_0 = 0,$	$\beta_1 = 1,$	$\beta_2 = 1,$	$\beta_3 = 0,$	
	$s_0 = 0,$	$s_1 = 1,$	$s_2 = 2,$	$s_3 = 1,$	
$\mathbf{c}_2 = 10^3 1010^3 1$	$b_0 = 0,$	$b_1 = 3,$	$b_2 = 1,$	$b_3 = 3,$	$b_4 = 0,$
	$\beta_0 = 0,$	$\beta_1 = 1,$	$\beta_2 = 1,$	$\beta_3 = 1,$	$\beta_4 = 0,$
	$s_0 = 0,$	$s_1 = 1,$	$s_2 = 2,$	$s_3 = 0,$	$s_4 = 0.$

Now, suppose $j \ge 1$. The jth run of zeros is preceded by a 1. Let the sequence obtained by changing this 1 into 0, be denoted by $\mathbf{c}'$. This sequence has one run of zeros less than $\mathbf{c}$ since the $(j-1)$th and jth runs now have been combined into one, and its parity is $\gamma \equiv (\beta_{j-1} + 1 + \beta_j) \pmod 2$. Let $s'_0, s'_1, \ldots, s'_{w-1}$ be the check of $\mathbf{c}'$. Then

$$\begin{aligned} s'_i &= s_i \text{ for } i \le j-2, \\ s'_{j-1} &\equiv ((2-\gamma)s_{j-2} + \gamma) \pmod 3, \\ s'_{i-1} &\equiv ((2-\beta_i)s'_{i-2} + \beta_i) \pmod 3 \text{ for } j \le i \le w. \end{aligned}$$

By (4.25)

$$s_j \equiv ((2-\beta_j)((2-\beta_{j-1})s_{j-2} + \beta_{j-1}) + \beta_j) \pmod 3.$$

The possible values of s_j and s'_{j-1} are given in Table 4.7.

Table 4.7 Possible values of s_j and s'_{j-1}.

β_j	β_{j-1}	γ	$s_j \equiv$	$s'_{j-1} \equiv$	$(s_j - s'_{j-1}) \pmod 3$
0	0	1	s_{j-2}	$s_{j-2}+1$	2
0	1	0	$2s_{j-2}+2$	$2s_{j-2}$	2
1	0	0	$2s_{j-2}+1$	$2s_{j-2}$	1
1	1	1	$s_{j-2}+2$	$s_{j-2}+1$	1

From Table 4.7 we see that

$$s_j - s'_{j-1} \equiv 2 - \beta_j \not\equiv 0 \pmod 3. \tag{4.26}$$

Since $2 - \beta_i \not\equiv 0 \pmod 3$ for all i, induction shows that $s_i \neq s'_{i-1}$ for $i = j, j+1, \ldots, w$. Hence, $s(\mathbf{c}) \neq s(\mathbf{c}')$.

Changing a 0 to a 1 is the opposite operation. Hence this will also change the check value.

Finally, consider shifts. Let $\mathbf{c}''$ denote the sequence obtained by shifting the 1 preceding the jth run of zeros one step to the right. Then $b''_{j-1} = b_{j-1} + 1$ and $b''_j = b_j - 1$. In particular, $\beta''_j = 1 - \beta_j$. If we remove this 1 from $\mathbf{c}''$ we again obtain $\mathbf{c}'$. Hence, by (4.26),

$$s''_j - s'_{j-1} \equiv 2 - (1 - \beta_j) = \beta_j + 1.$$

Combining this with (4.26), we get

$$s''_j - s_j \equiv 1 - 2\beta_j \not\equiv 0 \pmod 3.$$

Again, induction shows that $s(\mathbf{c}'') \neq s(\mathbf{c})$.

This shows that the set of all (d, k) constrained sequences of length n with a fixed check value is code that can detect any single bit-error or shift. Since we have three possible check values, this defines a partition of $R_{n,d,k}$ into three such codes.

Construction of a systematic code

Let $\mathbf{c}$ be a (d, k) constrained sequence. We first split this into parts $\mathbf{x}_i$ of length n, say, that is

$$\mathbf{c} = \mathbf{x}_1\mathbf{x}_2\mathbf{x}_3 \cdots .$$

We want to find sequences $\mathbf{y}_i$ for $i = 1, 2, 3, \ldots$ such that

- The concatenated sequence $\mathbf{x}_i\mathbf{y}_i$ has check value zero for all i.
- The concatenated sequence $\mathbf{x}_1\mathbf{y}_1\mathbf{x}_2\mathbf{y}_2\mathbf{x}_3\mathbf{y}_3 \cdots$ is (d, k) constrained.

Consider $\mathbf{x}_i$. Let u the length of the last run of zeros in $\mathbf{x}_i$ and v the length of the first run of zeros in $\mathbf{x}_{i+1}$, that is

$$\mathbf{x}_i = \cdots 10^u \text{ and } \mathbf{x}_{i+1} = 0^v 1 \cdots .$$

Then

$$u \geq 0,\ v \geq 0,\ d \leq u + v \leq k.$$

We first consider the simpler case $k = \infty$. It can be shown that $\mathbf{y}_i$ has to be of length at least $d+2$. We will describe how one can choose $\mathbf{y}_i$ of length $d+2$. If $u \le d$, we present three candidates for $\mathbf{y}_i$, namely

$$\mathbf{z}_1 = 0^{d-u}10^{u+1},\ \mathbf{z}_2 = 0^{d-u+1}10^{u},\ \mathbf{z}_3 = 0^{d+2}.$$

First we observe that all of them will make the concatenated sequence satisfy the (d,∞) constraint. We further observe that $\mathbf{z}_2$ is obtained from $\mathbf{z}_1$ by a right shift of the 1, and $\mathbf{z}_3$ is obtained from $\mathbf{z}_1$ by changing the 1 to 0. From the analysis of the non-systematic code above, this means that the three sequences $\mathbf{x}_i\mathbf{z}_1$, $\mathbf{x}_i\mathbf{z}_2$, and $\mathbf{x}_i\mathbf{z}_3$ have distinct check values. Hence, one of $\mathbf{z}_1$, $\mathbf{z}_2$, $\mathbf{z}_3$ can be chosen as $\mathbf{y}_i$ to get the check value zero.

If $u > d$, we again have three candidates for $\mathbf{y}_i$, namely

$$10^{d+1},\ 010^{d},\ 0^{d+2}.$$

The analysis is similar to the previous case.

For finite $k \ge 2d+3$, it can be shown that $\mathbf{y}_i$ has to be of length at least $2d+3$. We will describe how one can choose $\mathbf{y}_i$ of length $2d+3$. The description is now split into four cases which cover all possibilities for u and v. The analysis is again similar and we just give the three candidates for $\mathbf{y}_i$ in each case in Table 4.8.

Table 4.8 Three candidates for $\mathbf{y}_i$.

range	candidates		
$u \le d$	$0^{d-u}10^{d+1}10^{u}$	$0^{d-u+1}10^{d}10^{u}$	$0^{2d-u+2}10^{u}$
$v \le d < u$	$0^{v}10^{d}10^{d+1-v}$	$0^{v}10^{d+1}10^{d-v}$	$0^{v}10^{2d+2-v}$
$d < u \le \lfloor k/2 \rfloor,\ v > d$	$10^{d+1}10^{d}$	$010^{d}10^{d}$	$0^{d+2}10^{d}$
$d < v < \lfloor k/2 \rfloor < u$	$0^{d}10^{d}10$	$0^{d}10^{d+1}1$	$0^{d}10^{d+2}$

4.5 Comments and references

4.1. Theorem 4.2 is due to de Bruijn, van Ebbenhorst Tengbergen, and Kruyswijk (1951).

The systematic perfect code given is due to Berger (1961). Freiman (1962) proved that no perfect systematic code has less redundancy.

Theorem 4.1 is from Kløve and Korzhik (1995). Freiman suggested using the code with $\{a_1, a_2, \ldots a_k\} = \{\lfloor n/2 \rfloor \pm i(a+1) \mid 0 \le i \le n/(2(a+1))\}$ to detect up to a asymmetric errors. Note that Example 4.2 shows that this construction may not be optimal. Freiman also has a suggestion for a systematic code to detect up to a asymmetric errors. A channel related to the Z-channel is the *unidirectional channel.* For this channel, if there are errors in a code word, they are all $0 \to 1$ errors or all $1 \to 0$ errors. A number of codes have been constructed for correction or detecting of such error-patterns. Also there are many constructions of codes correcting t symmetric errors (where t usually is 1) and detecting all unidirectional errors. For more information on such codes, we refer the reader to the books by Rao and Fujiwara (1989) and Blaum (1993).

The Generalized Bose-Lin codes and their analysis were given by Gancheva and Kløve (2005b) who proved Theorems 4.3 and 4.4. For $q = 2$ (and ω even) we get the codes given by Bose and Lin (1985). For $\omega = 0$ (and general q) we get the codes given by Bose and Pradhan (1982) (actually, Bose and Pradhan considered the codes with the smallest possible value of r, namely $r = \lceil \log_q((q-1)k+1) \rceil$). For $\omega = 2$ (and general q) we get the codes studied by Bose, Elmougy, and Tallini (2005), El-Mougy (2005), El-Mougy and Gorshe (2005).

The presentation of diversity combining is based on the paper by Kløve, Oprisan, and Bose (2005a). Their paper contains a number of additional results not presented in this book.

4.2. The results are essentially due to Karpovsky and Taubin (2004) (with a different notation). See also Karpovsky and Nagvajara (1989).

4.3. Theorem 4.9 and the various constructions in Section 4.3.1 are taken from AbdelGhaffar (1998). This paper also contains a good bibliography on ST codes. Another recommendable survey was given by Gallian (1996).

Description of the various decimal codes and barcodes can be found many places on the web. As URLs often become obsolete, I do not include any here.

The Verhoeff code was given by Verhoeff (1969). The use of various algebraic structures, e.g. quasigroups, for construction of single check digits have been studied by a number of authors since. We refer to the paper by Belyavskaya, Izbash, and Mullen (2005a) which contains a good bibliography for such codes.

The Norwegian personal number codes were constructed by Selmer (1967).

4.4. The non-systematic code was given by Perry (1995) and the systematic code by Perry, Li, Lin, and Zhang (1998).

Bibliography

K.A.S. Abdel-Ghaffar, "A lower bound on the undetected probability of block codes", *Proc. 1994 IEEE Int. Symp. Inform. Theory* (1994) 341.

K.A.S. Abdel-Ghaffar, "A lower bound on the undetected error probability and strictly optimal codes", *IEEE Trans. Inform. Theory* 43 (1997) 1489–1502.

K.A.S. Abdel-Ghaffar, "Detecting substitutions and transpositions of characters", *The Computer Journal* 41(4) (1998) 270–277.

K.A.S Abdel-Ghaffar, "Repeated use of codes for error detection five times is bad", *IEEE Trans. Inform. Theory* 48 (2002) 2053–2060.

K.A.S Abdel-Ghaffar, "A simple derivation of the undetected error probabilties of complementary codes", *IEEE Trans. Inform. Theory* 50 (2004) 861–862.

K.A.S. Abdel-Ghaffar and H.C. Ferreira, "Systematic encoding of the Varshamov-Tenengol'ts codes and the Constantin-Rao codes", *IEEE Trans. Inform. Theory* 44 (1998) 340–345.

V. K. Agarwal and A. Ivanov, "Computing the probability of undetected error for shortened cyclic codes", *IEEE Trans. Commun.* 40 (1992) 494–499.

L. Ahlfors, *Complex Analysis*, McGraw-Hill Book Company, New York, 1953.

S. Al-Bassam and B. Bose, "Asymmetric/unidirectional error correcting and detecting codes", *IEEE Trans. Inform. Computers* 43 (1994) 590–597.

S. Al-Bassam, B. Bose, and R. Venkatesan, "Burst and unidirectional error detecting codes", *Proc. FTCS* (1991) 378–384.

R. Anand, K. Ramchandran, and I. V. Kozintsev, "Continuous error detection (CED) for reliable communication", *IEEE Trans. Commun.* 49 (2001) 1540–1549.

T. C. Ancheta, "An upper bound on the ratio of the probabilities of subgroups and cosets", *IEEE Trans. Inform. Theory* 27 (1981) 646–647.

A. Ashikhmin and A. Barg, "Binomial moments of the distance distribution: Bounds and applications", *IEEE Trans. Inform. Theory* 45 (1999) 438–452.

A. Ashikhmin, A. Barg, E. Knill, *et al.*, "Quantum error detection I: Statement of the problem", *IEEE Trans. Inform. Theory* 46 (2000) 778–788.

A. Ashikhmin, G. Cohen, M. Krivelevich, *et al.*, "Bounds on distance distributions in codes of known size", *IEEE Trans. Inform. Theory* 51 (2005) 250–258.

E. F. Assmus and J. K. Key, *Designs and Their Codes*, Cambridge Univ. Press, Cambridge, 1993.

E. F. Assmus and H. F. Mattson, "The weight distribution of a coset of a linear code", *IEEE Trans. Inform. Theory* 24 (1978) 497.

T. Baicheva, S. Dodunekov, and P. Kazakov, "On the cyclic redundancy-check codes with 8-bit redundancy", *Computer Commun.* 21 (1998) 1030–1033.

T. Baicheva, S. Dodunekov, and P. Kazakov, "Undetected error probability performance of cyclic redundancy-check codes of 16-bit redundancy", *IEE Proc. Commun.* 147 (2000) 253–256.

T. Baicheva, S. Dodunekov, and R. Koetter, "On the performance of the ternary [13,7,5] quadratic-residue code", *IEEE Trans. Inform. Theory* 48 (2002) 562–564.

A. Barg, "On some polynomials related to weight enumerators of linear codes", *SIAM Journal on Discrete Math.* 15 (2002) 155–164.

A. Barg and A. Ashikhmin, "Binomial moments of the distance distribution and the probability of undetected error", *Designs, Codes and Cryptography* 16 (1999) 103–116.

A. M. Barg and I. I. Dumer, "On computing the weight spectrum of cyclic codes", *IEEE Trans. Inform. Theory* 38 (1992) 1382–1386.

L. D. Baumert and R. J. McEliece, "Weights of irreducible cyclic codes", *Information and Control* 20 (1972) 158–175.

G. B. Belyavskaya, V. I. Izbash, G. L. Mullen, "Check character systems using quasigroups: I", *Designs, Codes and Cryptography* 37 (2005) 215–227.

G. B. Belyavskaya, V. I. Izbash, G. L. Mullen, "Check character systems using quasigroups: II", *Designs, Codes and Cryptography* 37 (2005) 405–419.

G. Benelli, "A new method for integration of modulation and channel coding in an ARQ protocol", *IEEE Trans. Commun.* 40 (1992) 1594–1606.

G. Benelli, "Some ARQ protocols with finite receiver buffer", *IEEE Trans. Commun.* 41 (1993) 513–523.

G. Benelli, "New ARQ protocols using concatenated codes", *IEEE Trans. Commun.* 41 (1993) 1013–1019.

G. Benelli, "A new selective ARQ protocol with finite length buffer", *IEEE Trans. Commun.* 41 (1993) 1102–1111.

G. Benelli and A. Garzelli, "Type-II hybrid ARQ protocol using concatenated codes", *IEE Proc.-I Commun., Speech and Vision* 140 (1993) 346–350.

R. J. Benice and A. H. Frey Jr., "An analysis of retransmission systems", *IEEE Trans. Commun.* 12 (1964) 135–145.

J. M. Berger, "A note on error detecting codes for asymmetric channels", *Inform. Control* 4 (1961) 68–73. Reprinted in Blaum (1993).

E. R. Berlekamp, R. J. McEliece, H. C. A. van Tilborg, "On the inherent intractability of certain coding problems", *IEEE Trans. Inform. Theory* 24 (1978) 384–386.

D. Bertsekas and R. Gallager, *Data networks*, Prentice-Hall, Inc., Englewood Cliffs, New Jersey, 1986.

R. T. Bilous and G. H. J. van Rees, "An enumeration of binary self-dual codes of length 32", *Designs, Codes and Cryptography* 26 (2002) 61–68.

G. Birkhoff and S. Mac Lane, *A Survey of Modern Algebra*, The MacMillan Company, New York, 1953.

M. Blaum, *Codes for detecting and correcting unidirectional errors*, IEEE Comp. Soc. Press, Los Alamitos, Cal. 1993.

M. Blaum and H. van Tilborg, "On T-error correcting all unidirectional detecting codes", *IEEE Trans. Computers* 38 (1989) 1493–1501.

V. Blinovsky, "New estimation of the probability of undetected error", *Proc. 1995 IEEE Int. Symp. Inform. Theory* (1995) 57.

V. Blinovsky, "Estimate for the Probability of Undetected Error", *Prob. Peredachi Inform.* 32 no.2 (1996) (in Russian) [English translation: *Prob. Inform. Transmission* 32 (1996) 139–144].

I. Blake and K. Kith, "On the complete weight enumerator of Reed-Solomon codes" *SIAM Journal on Discrete Math.* 4 (1991) 164–171.

M. Blaum, J. Bruck, and L. Tolhuizen, "A note on 'A systematic (12,8) code for correcting single errors and detecting adjacent error' ", *IEEE Trans. Computers* 43 (1994) 125.

I. E. Bocharova and B. D. Kudryashov, "Development of a discrete channel model for fading channels" *Prob. Peredachi Inform.* 29 no.1 (1993) 58–67 (in Russian) [English translation: *Prob. Inform. Transmission* 29 (1993) 50–57].

J. M. Borden, "Optimal asymmetric error detecting codes", *Inform. Control* 53 (1982) 66–73. Reprinted in Blaum (1993).

B. Bose and S. Al-Bassam, "Byte unidirectional error correcting and detecting Codes", *IEEE Trans. Computers* 41 (1992) 1601–1606.

B. Bose, "Burst Unidirectional Error-Detecting Codes", *IEEE Trans. Computers* 36 (1986) 350–353.

B. Bose, S. Elmougy, and L.G. Tallini, "Systematic t-unidirectional error-detecting codes in Z_m", manuscript, 2005.

B. Bose and D.J. Lin, "Systematic unidirectional error-detecting codes", *IEEE Trans. Comp.* 34 (1985) 1026–1032.

B. Bose and D.K. Pradhan, "Optimal unidirectional error detecting/correcting codes", *IEEE Trans. Comp.* 31 (1982) 564–568.

D. A. H. Brown, "Biquinary decimal error detection codes with one, 2 and 3 check digits", *Computer Journal* 17 (1974) 201–204.

P. Camion, B. Courteau, and A. Montpetit, "Weight distribution of cosets of 2-error-correcting binary BCH codes of length 15, 63, and 255", *IEEE Trans. Inform. Theory* 38 (1992) 1353–1357.

J. L. Carter and M. N. Wegman, "Universal classes of hash functions", *J. Comp. System Sci.* 18 (1979) 143–154.

G. Castagnoli, J. Ganz, and P. Graber, "Optimal cyclic redundancy-check codes with 16 bit redundancy", *IEEE Trans. Commun.* 38 (1990) 111–114.

G. Castagnoli, S. Bräuer, and M. Herrmann, "Optimization of cyclic redundancy-check codes with 24 and 32 parity bits", *IEEE Trans. Commun.* 41 (1993) 883–892.

E. H. Chang, "Theory of information feedback systems", *IRE Trans. Inform. Theory* 2, no. 3 (1956) 29–42.

E. H. Chang, "Improvement of two-way communication by means of feedback",

IRE Nat. Conv. Rec pt 4 (1966) 88–104.

P. Charpin, "Weight distribution of the cosets of the binary 2-error-correcting BCH codes", *C. R. Acad. Sci. I, Math.* 317, no. 10 (1994) 975–980.

D. Chun and J. K. Wolf, "Special hardware for computing the probability of undetected error for certain binary CRC codes and test results", *IEEE Trans. Commun.* 42 (1994) 2769–2772.

K. M. Cheung, "Identities and approximations for the weight distribution of Q-ary codes", *IEEE Trans. Inform. Theory* 36 (1990) 1149–1153.

G. C. Clark and J. B. Cain, *Error Correcting Codes for Digital Communication*, Plenum Press, New York 1981.

G. D. Cohen, L. Gargano, and U. Vaccaro, "Unidirectional error detecting codes", *Springar Lecture Notes in Computer Science* 514 (1991) 94–105.

B. K. Dass and S. Jain, "On a class of closed loop burst error detecting codes", *Intern. Journal of nonlinear sciences and numerical simulation* 2 (2001) 305–306.

M. A. de Boer, "Almost MDS codes", *Designs, Codes and Cryptography* 9 (1996) 143-155.

N. G. de Bruijn, C. van Ebbenhorst Tengbergen, and D. Kruyswijk, "On the set of divisors of a number", *Nieuw Archief voor Wiskunde* (2) 23 (1951) 191-193.

P. Delsarte, "Bounds for unrestricted codes, by linear programming", *Philips Res. Reports* 27 (1972) 272–289.

R. H. Deng, "Hybrid ARQ schemes for point-to-multipoint communication over nonstationary broadcast channels", *IEEE Trans. Commun.* 41 (1993) 1379–1387.

R. H. F. Denniston, "Some maximal arcs in finite projective planes", *J. Comb. Theory* 6 (1969) 317–319.

Y. Desaki, T. Fujiwara, and T. Kasami, "The weight distributions of extended binary primitive BCH codes of length 128", *IEEE Trans. Inform. Theory* 43 (1997) 1364–1371.

C.S. Ding, T. Kløve, and F. Sica, "Two classes of ternary codes and their weight distributions", *Discrete Applied Math.* 111 (2001) 37–53.

R. L. Dobrushin, "The mathematical problems of Shannon's theory of optimal information coding" (in Russian), *Prob. Peredachi Inform.* 10, 1961, 63–107.

S. M. Dodunekov and I. N. Landgev, "On near-MDS codes", Report LiTH-ISY-R-1563, Dept. of Elec. Eng., Linköping Univ. (1994).

R. Dodunekova, "The duals of MMD codes are proper for error detection", *IEEE Trans. Inform. Theory* 49 (2003) 2034–2038.

R. Dodunekova, "Extended binomial moments of a linear code and the undetected error probability", *Prob. Peredachi Informat.* 39, no. 3 (2003) 28–39 (in Russian) [English translation: *Problems Inform. Transmission* 39, no. 3 (2003) 255–265]

R. Dodunekova and S. Dodunekov, "Linear block codes for error detection", in *Proc. 5th Intern. Workshop on Algebraic and Combinatorial Coding Theory, Sozopol* (1996) 117–122.

R. Dodunekova and S. Dodunekov, "Sufficient conditions for good and proper error-detecting codes via their duals", *Math. Balkanica (N.S.)* 11, Fasc. 3–4 (1997) 375–381.

R. Dodunekova and S. Dodunekov, "Sufficient conditions for good and proper error detecting codes", *IEEE Trans. Inform. Theory* 43 (1997) 2023–2026.

R. Dodunekova and S. Dodunekov, "Sufficient conditions for good and proper error correcting codes", in *Proc. 2th Inter. Workshop on Optimal Codes and Related topics, Sozopol* (1998) 62–67.

R. Dodunekova and S. Dodunekov, "The MMD codes are proper for error detection", *IEEE Trans. Inform. Theory* 48 (2002) 3109–3111.

R. Dodunekova and S. Dodunekov, "t-good and t-proper linear error correcting codes",
emphMathematica Balkanica (N.S.) 17, Fasc. 12 (2003) 147–154.

R. Dodunekova and S. Dodunekov, "Error detection with a class of cyclic codes", in *Proc. 4th Inter. Workshop on Optimal Codes and Related topics, Pamporovo* (2005) 127–132.

R. Dodunekova and S. Dodunekov, "Error detection with a class of q-ary two-weight codes", *Proc. IEEE International Symposium on Information Theory* (2005) 2232–2235.

R. Dodunekova, S. Dodunekov, and T. Kløve, "Almost-MDS and near-MDS codes for error detection", *IEEE Trans. Inform. Theory* 43 (1997) 285–290.

R. Dodunekova, S. Dodunekov, and E. Nikolova, "On the error-detecting performance of some classes of block codes", in *Annual Workshop on coding theory and applications, Bankya* (2003).

R. Dodunekova, S. Dodunekov, and E. Nikolova, "On the error-detecting performance of some classes of block codes", *Prob. Peredachi Informat.* 40, no. 4 (2004) 68–78 [English translation: *Problems Inform. Transmission* 40, no. 4 (2004) 356–364].

R. Dodunekova, S. Dodunekov, and E. Nikolova, "Non-linear binary codes for error detection", in *Proc. 9th Intern. Workshop on Algebraic and Combinatorial Coding Theory, Kranevo* (2004) 125–130.

R. Dodunekova, S. Dodunekov, and E. Nikolova, "A survey on proper codes", in General Theory of Information Transfer and Combinatorics, a special issue of *Discrete Applied Math.*, to appear.

R. Dodunekova and E. Nikolova, "Sufficient conditions for the monotonicity of the undetected error probability for large channel error probabilties", *Prob. Peredachi Informat.* 41, no. 3 (2005) 3–16 [English translation: *Problems Inform. Transmission* 41, no. 3 (2005) 187–198].

R. Dodunekova and E. Nikolova, "Properness of binary linear error-detecting codes in terms of basic parameters", in *Proc. 4th Intern. Workshop on Optimal Codes and Related topics, Pamporovo* (2005) 133–138.

R. Dodunekova, O. Rabaste, and J.L.V. Paez, "On the error-detecting performance of a class of irreducible binary cyclic codes and their duals", in *Proc. 9th Intern. Workshop on Algebraic and Combinatorial Coding Theory, Kranevo* (2004) 131–136.

R. Dodunekova, O. Rabaste, and J.L.V. Paez, "Error detection with a class of irreducible binary cyclic codes and their dual codes", *IEEE Trans. Inform. Theory* 51 (2005) 1206–1209.

H. B. Dwight, *Tables of Integrals and other Mathematical Data*, MacMillan Co., New York, 1961.

S. El-Mougy, "Some contributions to asymmetric error control codes", PhD thesis, Oregon State Univ., 2005.

S. El-Mougy and S. Gorshe, "Some error detecting properties of Bose-Lin codes", submitted to *IEEE Trans. on Computers* 2005.

T. Etzion, "Optimal codes for correcting single errors and detecting adjacent errors", *IEEE Trans. Inform. Theory* 38 (1992) 1357–1360.

A. Faldum, "Trustworthiness of error-correcting codes", manuscript (2005).

A. Faldum and K. Pommerening, "An optimal code for patient identifiers", *Computer Mathods and Programs in Biomedicine* 79 (2005) 81–88.

A. Faldum, J. Lafuente, G. Ochoa, and W. Willems, "Error Probabilities for Bounded Distance Decoding", *Designs, Codes and Cryptography* 40 (2006) 237–252.

A. Faldum and W. Willems, "Codes of small defect", *Designs, Codes and Cryptography* 10 (1997) 341-350.

A. Faldum and W. Willems, "A characterization of MMD codes", *IEEE Trans. Inform. Theory* 44 (1998) 1555–1558.

R. Fantacci, "Performance evaluation of some efficient stop-and-wait techniques", *IEEE Trans. Commun.* 40 (1992) 1665–1669.

P. Farkas, "Weighted sum codes for error-detection and their comparison with existing codes - comments", *IEEE-ACM Trans. on Networking* 3 (1995) 222–223.

D. C. Feldmeier, "Fast software implementation of error detection codes", *IEEE-ACM Trans. on Networking* 3 (1995) 640–651.

W. Feller, *An introduction to probability theory and its applications*, 2nd ed., vol.1, John Wiley & Sons, Inc., London 1957.

L. M. Fink, *Theory of Digital Communication* (in Russian), Sov. Radio, Moscow 1965.

L. M. Fink and V. I. Korzhik, "Selection of codes for error detection in decision-feedback systems" (in Russian), in *Proc. Third Int. Symp. Inform. Theory* no. 1 (1973) 138–142.

L. M. Fink and S. A. Mukhametshina, "Pseudostochastic coding for error detection", *Prob. Peredachi Inform.* 15, no. 2 (1979) 36–39 (in Russian) [English translation: *Prob. Inform. Transmission* 15 (1979) 108–110].

S. Fratini, "Error detection in a class of decimal codes", *IEEE Trans. Inform. Theory* 35 (1989) 1095–1098.

C. V. Freiman, "Optimal error detection codes for completely asymmetric channels", *Inform. Control* 5 (1962) 64–71. Reprinted in Blaum (1993).

F. W. Fu and T. Kløve, "Large binary codes for error detection", *Reports in Informatics* no. 315, Department of Informatics, Univ. of Bergen, (2006).

F. W. Fu, T. Kløve, and V.K.W. Wei, "On the undetected error probability for binary codes", *IEEE Trans. Inform. Theory* 49 (2003) 382–390.

F. W. Fu, T. Kløve, and S. T. Xia, "On the Undetected Error Probability of

m-out-of-n Codes on the Binary Symmetric Channel", in J. Buchmann, T. Høholdt, H. Stichtenoth, H. Tapia-Recillas (Eds.): *Coding Theory, Cryptography, and Related Areas*, Springer, 2000, 102–110. (Proc. Intern. Conf. on Coding Theory, Cryptography and Related Areas, Guanajuato, Mexico, 20–24 April, 1998).

F. W. Fu, T. Kløve, and S. T. Xia, "The undetected error probability threshold of m-out-of-n codes", *IEEE Trans. Inform. Theory* 46 (2000) 1597–1599.

F. W. Fu and S. T. Xia, "Binary constant-weight codes for error detection", *IEEE Trans. Inform. Theory* 44 (1998) 1294–1299.

E. Fujiwara and M. Sakura, "Nonsystematic d-unidirectional error detecting codes", *Proc. IEEE Int. Symp. Inform. Theory* (1990) 174.

T. Fujiwara and T. Kasami, "Probability of undetected error after decoding for a concatenated coding scheme", *Proc. IEEE Int. Symp. Inform. Theory* (1985) 110.

T. Fujiwara and T. Kasami, "Error detecting capabilities of the shortened Hamming codes adopted for error detection in IEEE standard 802.3", *Proc. IEEE Int. Symp. Inform. Theory* (1986) 65.

T. Fujiwara, T. Kasami, and S. Feng, "On the monotonic property of the probability of undetected error for a shortened code", *IEEE Trans. Inform. Theory* 37 (1991) 1409–1411.

T. Fujiwara, T. Kasami, A. Kitai, and S. Lin, "On the undetected error probability for shortened Hamming codes", *IEEE Trans. Commun.* 33 (1985) 570–574.

T. Fujiwara, T. Kasami, and S. Lin, "Error detecting capabilities of the shortened Hamming codes adopted for error detection in IEEE standard 802.3", *IEEE Trans. Commun.* 37 (1989) 986–989.

T. Fujiwara, A. Kitai, S. Yamamura, T. Kasami, and S. Lin, "On the undetected error probability for shortened cyclic Hamming codes", in *Proc. 5th Conf. Inform. Theory and Its Appl.*, Hachimantai, Japan, Oct. 1982.

G. Funk, "Determination of best shortened linear codes", *IEEE Trans. Commun.* 44 (1996) 1–6.

G. Funk, "Message error detection properties of HDLC protocols", *IEEE Trans. Commun.* 30 (1982) 252–257.

J. A. Gallian, "Error detetion methods", *ACM Comput. Surveys* 28 (1996) 504–517.

I. Gancheva and T. Kløve, "Constructions of some optimal t-EC-AUED codes", *Proceedings, Optimal Codes and Related Topics,* Pamporovo, Bulgaria, June 17-23 (2005) 152-156.

I. Gancheva and T. Kløve, "Generalized Bose-Lin codes, a class of codes detecting asymmetric errors", *Proceedings, Optimal Codes and Related Topics,* Pamporovo, Bulgaria, June 17-23 (2005) 157-162.

I. Gancheva and T. Kløve, "Codes for error detection, good or not good", *Proc. IEEE Intern. Symp. on Inform. Th.* (2005) 2228–2231.

S. W. Golomb, "A probability calculation and estimate for Reed-Solomon codes", unpublished.

T. A. Gulliver and V. K. Bhargava, "A systematic (16,8) code for correcting double errors and detecting triple-adjacent errors", *IEEE Trans. on Computers* 42 (1993) 109–112.

T. Hashimoto, "Good error-detection codes satisfying the expurgated bound", *IEEE Trans. Inform. Theory* 41 (1995) 1347–1353. cfr. Hashimoto (1994)

T. Hashimoto, "Composite scheme LR+Th for decoding with erasures and its effective equivalence to Forney's rule", *IEEE Trans. Inform. Theory* 45 (1999) 78–93.

T. Hashimoto and M. Taguchi, "Performance of explicit error detection and threshold decision in decoding with erasures", *IEEE Trans. Inform. Theory* 43 (1997) 1650–1655.

Y. Hayashida, "Throughput analysis of tandem-type go-back-N ARQ scheme for satellite communication", *IEEE Trans. Commun.* 41 (1993) 1517–1524.

A. R. Hammons, P. V. Kumar, A. R. Calderbank, N. J. A. Sloane, and P. Solé, "The Z_4-linearity of Kerdock, Preparata, Goethals and related codes", *IEEE Trans. Inform. Theory* 40 (1994) 301–319.

G. H. Hardy and E. M. Wright, *An Introduction to the Theory of Numbers*, Oxford Univ. Press 1960.

B. C. Harvey and S. B. Wicker, "Packet combining systems based on the Viterbi decoder", *IEEE Trans. Commun.* 42 (1994) 1544–1557.

T. Hashimoto, "Good error detection codes satisfy the expurgated bound", *Proc. IEEE Int. Symp. Inform. Theory* (1994) 343.

R. S. He and J. R. Cruz "High-rate coding system for magnetic recording", *IEEE Trans. Magnetics* 35 (1999) 4522–4527.

T. Helleseth, T. Kløve, V. Levenshtein, Ø. Ytrehus, "Bounds on the minimum support weights", *IEEE Trans. Inform. Theory* 41 (1995) 432–440.

T. Helleseth, T. Kløve, and J. Mykkeltveit, "The weight distribution of irreducible cyclic codes", *Discrete Math.* 18 (1977) 179–211.

T. Helleseth, T. Kløve, Ø. Ytrehus, "Generalized Hamming weights of linear codes", *IEEE Trans. Inform. Theory* 38 (1992) 1133–1140.

T. Helleseth, T. Kløve, Ø. Ytrehus, "On generalizations of the Griesmer bound", *Reports in informatics*, no. 87, Department of Informatics, University of Bergen, September 1993.

M. E. Hellman, "Error detection made simple", in *Proc. IEEE Int. Conf. on Commun.*, Minneapolis, June 17–19 1974, 9A1–9A4.

I.S. Honkala and T.K. Laihonen, "The probability of undetected error can have several local maxima", *IEEE Trans. Inform. Theory* 45 (1999) 2537–2539.

W. H. Huggins, "Signal-flow graphs and random signals", *Proc. IRE* January (1957) 74–86.

Z. Huntoon and A. M. Michelson, "On the computation of the probability of post-decoding error events for block codes", *IEEE Trans. Inform. Theory* 23 (1977) 399–403.

H.K. Hwang, "Uniform asymptotics of some Abel sums arising in coding theory", *Theoretical Computer Science* 263 (2001) 145–158.

K. Imamura, K. Tokiwa, and M. Kasahara, "On computation of the binary weight distribution of some Reed-Solomon codes and their extended codes", *Proc. Int. Colloq. Coding Theory*, Osaka, Japan (1988) 195–204.

E. Isaacson and H. B. Keller, *Analysis of Numerical Methods*, John Wiley & Sons, Inc., New York, 1966.

I. Jacobs, "Optimum error detections codes for noiseless decision feedback", *IRE Trans. Inform. Theory* 8 (1962) 369–371.

R. Johansson, "A class of (12,8) codes for correcting single errors and detecting double errors within a nibble", *IEEE Trans. Computers* 42 (1993) 1504–1506.

M. G. Karpovsky and P. Nagvajara, "Optimal codes for minimax criterion on error detection", *IEEE Trans. Inform. Theory* 35 (1989) 1299–1305.

M. Karpovsky and A. Taubin, "New class of nonlinear systematic error detecting codes", *IEEE Trans. Inform. Theory* 50 (2004) 1818–1820.

T. Kasami, "The weight enumerators for several classes of subcodes of the 2nd order binary Reed-Muller codes", *Inform. Control* 18 (1971) 369–394.

T. Kasami, T. Fujiwara, and S. Lin, "Generalized Sidel'nikov's bound and its applications", *Proc. IEEE Int. Symp. Inform. Theory* (1983) 120.

T. Kasami, T. Fujiwara, and S. Lin, "An approximation to the weight distribution of binary linear codes", *IEEE Trans. Inform. Theory* 31 (1985).

T. Kasami, T. Fujiwara, and S. Lin, "A concatenated coding scheme for error control", *IEEE Trans. Commun.* 34 (1986) 481–488.

T. Kasami, T. Kløve, and S. Lin, "Linear block codes for error detection", *IEEE Trans. Inform. Theory* 29 (1983) 131–136.

T. Kasami and S. Lin, "On the probability of undetected error for the maximum distance separable codes", *IEEE Trans. Commun.* 32 (1984) 998–1006.

T. Kasami and S. Lin, "The binary weight distribution of the extended $(2^m, 2^m - 4)$ code of the Reed-Solomon codes over $GF(2^m)$ with generator polynomial $(x-\alpha)(x-\alpha^2)(x-\alpha^3)$", *Linear Algebra and Its Applications* 98 (1988) 291–307.

G. L. Katsman, "Upper bounds on the probability of undetected error", *Proc. Fourth Intern. Workshop on Algebraic and Comb. Coding Th.*, Novgorod, Russia, Sept. 11–17 (1994).

P. Kazakov, "Fast calculation of the number of minimum-weight words of CRC codes", *IEEE Trans. Inform. Theory* 47 (2001) 1190–1195.

S. R. Kim and C. K. Un, "Throughput analysis for two ARQ schemes using combined transition matrix", *IEEE Trans. Commun.* 40 (1992) 1679–1683.

T. Kløve, "The probability of undetected error when a code is used for error correction and detection", *IEEE Trans. Inform. Theory* 30 (1984) 388–392.

T. Kløve, "Generalizations of the Korzhik bound", *IEEE Trans. Inform. Theory* 30 (1984) 771–773.

T. Kløve, "Using codes for error correction and detection", *IEEE Trans. Inform. Theory* 30 (1984) 868–871.

T. Kløve, "Codes for error correction and detection" *Proc. 6'th Int. Symp. on Information Theory*, Tashkent, Uzbekistan, Sept. 18-22, vol. 2 (1984) 228-230.

T. Kløve, "Codes for error detection", unpublished lecture notes, presented in the Nordic Summer School on Coding Theory, Bergen 1987.

T. Kløve, "Optimal codes for error detection", *IEEE Trans. Inform. Theory* 38

(1992) 479–489.

T. Kløve, "A lower bound on the probability of undetected error", *Proc. Sixth Joint Swedish-Russian Int. Workshop on Inform. Theory*, Mölle (1993) 362–366.

T. Kløve, "On Massey's bound on the worst-case probability of undetected error", *Proc. IEEE Int. Symp. Inform. Theory* (1994) 242.

T. Kløve, "The weight distribution of cosets", *IEEE Trans. Inform. Theory* 40 (1994) 911–913.

T. Kløve, "Bounds on the worst-case probability of undetected error", *IEEE Trans. Inf. Theory* 41 (1995) 298–300.

T. Kløve, "Near-MDS codes for error detection", *Proc. Int. Workshop on Optimal Codes and Related Topics*, Sozopol, Bulgaria, 26 May-1 June 1995, 103–107.

T. Kløve, "The worst-case probability of undetected error for linear codes on the local binomial channel", *IEEE Trans. Inform. Theory* 42 (1996) 172–179.

T. Kløve, "Reed-Muller codes for error detection: The good, the bad, and the ugly", *IEEE Trans. Inform. Theory* 42 (1996) 1615–1622.

T. Kløve, "Bounds on the weight distribution of cosets", *IEEE Trans. Inform. Theory* 42 (1996) 2257–2260.

T. Kløve and V. I. Korzhik, *Error detecting Codes, General Theory and Their Application in Feedback Communication Systems*, Kluwer Acad. Publ., Boston 1995.

T. Kløve and M. Miller, "The detection of error after error-correction decoding", *IEEE Trans. on Commun.* 32 (1984) 511–517.

T. Kløve, P. Oprisan, and B. Bose, "Diversity combining for the Z-channel", *IEEE Trans. Inform. Theory* 51 (2005) 1174–1178.

T. Kløve, P. Oprisan, and B. Bose, "The probability of undetected error for a class of asymmetric error detecting codes", *IEEE Trans. Inform. Theory* 51 (2005) 1202–1205.

E. Kolev and N. Manev, "The binary weight distribution of an extended $(2M, 5)$ Reed-Solomon code and its dual", *Doklady Bolgarskoi Akademii Nauk* 43, Iss. 9 (1990) 9–12.

E. Kolev and N. Manev, "The binary weight distribution of the extended $[2^m, 5]$ Reed-Solomon code and its dual code" (in Russian), *Prob. Peredach. Inform.* 30, no. 3 (1994) (in Russian) [English translation: *Prob. Inform. Transmission*].

V. I. Korzhik, "Bounds on undetected error probability and optimum group codes in a channel with feedback", *Radioteckhnika* 20, vol. 1 (1965) 27–33 (in Russian) [English translation: *Telecommun. Radio Eng.* 20, no.1, part 2 (1965) 87–92].

V. I. Korzhik and S. D. Dzubanov, "Error detection codes for local binomial channels", *Proc. The Third Int. Colloquium on Coding Theory*, Dilijan, Armenia, Sept. 25 - Okt. 2 (1990) 127–130.

V. I. Korzhik and L. M. Fink, *Noise-Stable Coding of Discrete Messages in Channels with a Random Structure*, (in Russian), Svyaz, Moscow 1975.

V. I. Korzhik and L. M. Fink, "The multistage stochastic coding", *Int. Symp. Inform. Theory*, Repino, USSR (1976).

V. I. Korzhik, M. Y. Lesman and P. Malev, "An ARQ scheme using concatenated codes for mobile satellite channels and its throughput efficiency optimization", *Proc. Int. Conf. Russat '93*, St. Petersburg (1993) 39–42.

V. I. Korzhik, S. A. Osmolovskii, and L. M. Fink, "Universal stochastic coding in decision-feedback systems", *Prob. Peredachi Inform.* 10, no. 4 (1974) 25–29 (in Russian) [English translation: *Prob. Inform. Transmission* 10 (1974) 296–299].

V. I. Korzhik and V. A. Yakovlev, "Nonasymptotic estimates of information protection efficiency for the wire-tap channel concept", Auscrypt 92, *Springer Lecture Notes in Computer Science* 718 (1992) 185–195.

M. A. Kousa and M. Rahman, "An adaptive error control-system using hybrid ARQ schemes", *IEEE Trans. Commun.* 39 (1991) 1049–1057.

H. Krishna and S. D. Morgera, "A new error control scheme for hybrid ARQ systems", *IEEE Trans. on Commun.* 35 (1987) 981–990.

P. V. Kumar and V. Wei, "Minimum distance of logarithmic and fractional partial m-sequences", *IEEE Trans. Inform. Theory* 38 (1992) 1474–1482.

S. Kundu and S. M. Reddy, "On symmetric error correcting and all unidirectional error detecting codes", *Proc. IEEE Int. Symp. Inform. Theory* (1988) 120.

A. Kuznetsov, F. Swarts and H. C. Ferreira, "On the undetected error probability of linear block codes on channels with memory", *Proc. Sixth Joint Swedish-Russian Int. Workshop on Inform. Theory*, Mölle (1993) 367–371.

A. Kuznetsov, F. Swarts, A. J. H. Vinck, and H. C. Ferreira, "On the undetected error probability of linear block codes on channels with memory", *IEEE Trans. Inform. Theory* 42 (1996) 303–309.

J. N. Laneman, C. E. W. Sundberg, and C. Faller, "Huffman code based error screening and channel code optimization for error concealment in Perceptual Audio Coding (PAC) algorithms", *IEEE Trans. Broadcasting.* 48 (2002) 193–206.

J. Lawson, A. Michelson, and Z. Huntoon, "A performance analysis of a family of codes constructed from simple binary codes", presented at *Canadian Conf. Commun. and Power*, Montreal, PQ, Canada, Oct. 1978.

C. F. Leanderson and C. E. W. Sundberg, "Performance evaluation of list sequence MAP decoding", *IEEE Trans. Commun.* 53 (2005) 422–432.

V. K. Leontev, "Asymptotic optimum group codes in channels with feedback", presented at *All-Union Conf. on Probl. of Theoretical Cybernetics,* Novosibirsk 1969 (in Russian).

V. K. Leontev, "Error-detecting encoding", *Prob. Peredach. Inform.* 8, no. 2 (1972) 6–14 (in Russian) [English translation: *Prob. Inform. Transmission* 8 (1972) 86–92].

S. K. Leung, "Evaluation of the undetected error probability of single parity-check product codes", *IEEE Trans. Commun.* 31 (1983) 250–253.

S. K. Leung and M. E. Hellman, "Concerning a bound on undetected error probability", *IEEE Trans. Inform. Theory* 22 (1976) 235–237.

S. K. Leung, E. R. Barnes, and D. U. Friedman, "On some properties of the undetected error probability of linear codes", *IEEE Trans. Inform. Theory* 25 (1979) 110–112.

C. Leung and K. A. Witzke, "On testing for improper error detection codes", *IEEE Trans. Commun.* 38 (1990) 2085–2086.

V. I. Levenshtein, "Bounds on the probability of undetected error", *Prob. Peredachi Inform.* 13, no.1 (1977) 3–18 (in Russian) [English translation: *Prob. Inform. Transmission* 13 (1978) 1–12].

V. I. Levenshtein, "Straight-line bound for the undetected error exponent", *Prob. Peredachi Inform.* 25 no.1 (1989) 33–37 (in Russian) [English translation: *Prob. Inform. Transmission* 25 (1989) 24–27].

M.-C. Lin, "Bounds on the undetected error probabilities of linear codes for single-error correction and error detection", *Electronic Letters* 27 (1991) 2264–2265.

M.-C. Lin, "Undetected error probabilities of codes for both error correction and detection", *IEEE Trans. Inform. Theory* 36 (1990) 1139–1141.

M. Ch. Lin and M. Y. Gun, "The performance analysis of a concatenated ARQ scheme using parity retransmissions", *IEEE Trans. Commun.* 39 (1991) 1869–1874.

R.-D. Lin and W.-S. Chen "Fast calculation algorithm of the undetected errors probability of CRC codes", *Proc. Advanced Information Networking and Applications, AINA 2005*, vol. 2, pp. 480- 483.

S. Lin and D. J. Costello, *Error Control Coding, Fundamentals and Applications*, 2nd. ed., Prentice-Hall Publ, Englewood Cliffs, New Jersey, 2004.

S. Litsyn, "New upper bounds on error exponents", *IEEE Trans. Inform. Theory* 45 (1999) 385–398.

X. Liu, P. Farrell, and C. Boyd, "A unified code", *Lecture Notes in Computer Science* 1746 (1999) 84–93.

D.-L. Lu and J.-F. Chang, "The effect of return channel errors on the performance of ARQ protocols", *Proc. IEEE Int. Symp. Inform. Theory* (1991) 221.

D.-L. Lu and J.-F. Chang, "Performance of ARQ protocols in bursty channel errors", *Proc. IEEE Int. Symp. Inform. Theory* (1994) 375.

P.A. MacMahon, *Combinatory Analysis*, vol. I & II, Chelsea Publ. Co, New York, 1960.

J. MacWilliams, "A theorem on the distribution of weights in a systematic code", *Bell Syst. Tech. J.* 42 (Jan. 1963) 79–94.

J. MacWilliams and N. J. A. Sloane, *The theory of Error-Correcting Codes*, North-Holland Publishing Company, Amsterdam 1977.

P. Malev, *Investigation and Design of Effective Methods for Digital Communication with ARQ* (in Russian), PhD Thesis, LEIC, St. Petersburg, Russia (1993).

C. L. Mallow, V. Pless, and N. J. A. Sloane, "Self-dual codes over GF(3)", *SIAM J. Applied Math.* 31 (1976) 649–666.

J. Massey, "Coding techniques for digital data networks" in *Proc. Int. Conf. Inform. Theory Syst.*, NTG-Fachbeirichte 65, Berlin, Germany, Sept. 18–20 1978, 307–315.

A. J. Mcauley, "Weighted sum codes for error-detection and their comparison with existing codes", *IEEE-ACM Trans. Networking* 2 (1994) 16.

R. J. McEliece and H. Rumsey Jr., "Euler products, cyclotomy, and coding", *J.*

Number Theory 4 (1972) 302–311.

A. T. Memişoğlu and S. Bilgenr, "An adaptive hybrid ARQ scheme based on exponentially weighted moving averages for slowly time-varying communication channels", *Proc. IEEE Int. Symp. Inform. Theory* (1994) 377.

M. J. Miller, "The probability of undetected error for shortened cyclic codes", *Proc. IEEE Int. Symp. Inform. Theory* (1985) 111.

M. J. Miller and S. Lin, "Codes for error detection in data transmission and storage", *J. of Electrical and Electronics Eng., Australia* 3 (Sept 1986).

M. J. Miller, M. G. Wheal, G. J. Stevens, and A. Mezhvinsky, "The X.25 error detection code is a poor choice" *Proc. of the Institution of radio and Electronic Engineers Australia IREECON '85*, Sydney, September 1985.

M. J. Miller, M. G. Wheal, G. J. Stevens, and S. Lin, "The reliability of error detection schemes in data transmission", *J. of Electrical and Electronics Eng., Australia* 6 (June 1986) 123–131.

R. Morelos-Zaragoza, *The Art of Error Correcting Coding*, 2nd. ed., John Wiley & Sons, 2006.

I. Naydenova and T. Kløve, "Necessary conditions for codes to be good for error detection", *Abstracts, Annual Workshop Coding Theory and Applications*, Bankya, Bulgaria, Dec. 15-18, (2005) 24.

I. Naydenova and T. Kløve, "Necessary conditions for codes to be good for error detection", manuscript submitted for publication.

I. Naydenova and T. Kløve, "Large proper codes for error detection", *Proc. IEEE Information Theory Workshop*, Chengdu, China (2006) 170–174.

I. Naydenova and T. Kløve, "A bound on q-ary t-EC-AUED codes and constructions og some optimal ternary t-EC-AUED codes", *Proceedings. 2006 Int. Symposium on Information Theory and its applications,* Seoul, Korea, (2006) 136-139.

I. Naydenova and T. Kløve, "Generalized Bose-Lin codes, a class of codes detecting asymmetric errors", *IEEE Trans. Inform. Theory* 53 (2007) to appear.

G.D. Nguyen, "Error-detection codes: Algorithms and fast implementation", *IEEE Trans. Computers* 54 (2005) 1–11.

E. P. Nikolova, "A sufficient condition for properness of a linear error-detecting code and its dual", *Mathematics and Mathematical Education. Proc. 34th Spring Conf. of the Union of Bulgarian Mathematicians,* Borovets, Bulgaria, April 6–9 (2005) 136–139.

E. Nikolova, "Some binary codes are better for error detection than the 'average code' ", *Annual Workshop, Coding Theory and Applications,* Bankya, Bulgaria, December 15–18 (2005) 25.

T. Nishijima, "An upper bound on the average probability of an undetected error for the ensamble of generalized Reed-Solomon codes", *IEICE Trans. Fundamentals* J85-A, no. 1 (2002) 137–140.

T. Nishijima, "An upper and a lower bound on the probability of an undetected error for binary expansions of generalized Reed-Solomon codes", *Proc. IEEE Information Theory Workshop, Chengdu* (2006) 37–41.

T. Nishijima, O. Nagata, and S. Hirasawa, "On the probability of undetected error for iterated codes", *Proc. IEEE Int. Symp. Inform. Theory* (1993)

162.

T. Nishijima and S. Hirasawa, "On the probability of undetected error and the computational complexity to detect an error for iterated codes", *Proc. IEEE Int. Symp. Inform. Theory* (2005) 50.

V. K. Oduol and S. D. Morgera, "Performance evaluation of the generalized type-II hybrid ARQ scheme with noisy feedback on Markov channels", *IEEE Trans. Commun.* 41 (1993) 32–40.

J. Olsson and W. Willems, "A characterization of certain Griesmer codes: MMD codes in a more general sense", *IEEE Trans. Inform. Theory* 45 (1999) 2138–2142.

C. T. Ong and C. Leung, "On the undetected error probability of tripel-error-correcting BCH codes", *IEEE Trans. Inform. Theory* 37 (1991) 673–678.

Ø. Ore, *Theory of Graphs*, Amer. Math. Soc., Providence 1962

R. Padovani and J. K. Wolf, "Data transmission using error detection codes", in GLOBECOM '82, IEEE Global Telecom. Conf., Miami, Nov. 29– Dec. 2, 1982, 626–631.

R. Padovani and J. K. Wolf, "Poor error correction codes are poor error detection codes", *IEEE Trans. Inform. Theory* 30 (1984) 110–111.

J. Park and J. Moon, "Imposing a k constraint in recording systems employing post-Viterbi error correction", *IEEE Trans. Magnetics* 41 (2005) 2995–2997.

A. Patapoutian, B.Z. Shen, and P.A. McEwen, "Event error control codes and their applications", *IEEE Trans. Inform. Theory* 47 (2001) 2595–2603.

P. Perry, "Necessary conditions for good error detection", *IEEE Trans. Inform. Theory* 37 (1991) 375–378.

P. Perry, "Runlength-limited codes for single error detection in magnetic recording channel", *IEEE Trans. Inform. Theory* 41 (1995) 809–815.

P. Perry and M. Fossorier, "The expected value for the probability of an undetected error", *Proc. IEEE Information Theory Workshop, Chengdu* (2006) 219–223.

P. Perry and M. Fossorier, "The average probability of an undetected error", manuscript, 2006.

P. N. Perry, M. C. Li, M. C. Lin, and Z. Zhang, "Runlength limited codes for single error-detection and single error-correction with mixed type errors", *IEEE Trans. Inform. Theory* 44 (1998) 1588–1592.

W. W. Peterson, *Error-Correcting Codes*, MIT Press, Cambridge, Mass. 1961.

W. W. Peterson and E. J. Weldon, *Error-Correcting Codes*, MIT Press, Cambridge, Mass. 1972.

P. E. D. Pinto, F. Protti, and J. L. Szwarcfiter, "A Huffman-based error detecting code", *Lecture Notes in Computer Science* 3059 (2004) 446–457.

V. S. Pless, "Power moment identities on weight distributions in error correcting codes", *Information and Control* 6 (1963) 147–152.

V. S. Pless and W. C. Huffman, *Handbook of Coding Theory* I & II, Elsevier, Amsterdam 1998.

A. Poli and Ll. Huguet, *Error Correcting Codes, Theory and Applications*, Prentice Hall UK, Hemel Hempstead 1992.

D. K. Pradhan, "A new class of error-correcting/detecting codes for fault-tolerant computer applications", *IEEE Trans. Computers* 29 (1980) 471–481.

D. K. Pradhan and S. K. Gupta, "A new framework for designing and analyzing BIST techniques and zero aliasing compression", *IEEE Trans. Computers* 40 (1991) 743–763.

D. K. Pradhan, S. K. Gupta, and M. G. Karpovsky, "Aliasing probability for multiple input signature analyzer", *IEEE Trans. Computers* 39 (1990) 586–591.

J. Proakis, *Digital Communication*, McGraw-Hill Book Company, New York, 1989.

I. Rahman, "Nonbinary error detection codes for data retransmission and bit error rate monitoring", *IEEE Trans. Commun.* 40 (1992) 1139–1143.

T.R.N. Rao and E. Fujiwara, *Error-control coding for computer systems*, Prentice Hall, Inc., Englewood Cliffs, New Jersey, 1989.

L. K. Rasmussen and S. B. Wicker, "The performance of type-I, trellis coded hybrid-ARQ protocols over AWGN and slowly fading channels", *IEEE Trans. Inform. Theory* 40 (1994) 418–428.

L. K. Rasmussen and S. B. Wicker, "Trellis coded, type-I hybrid-ARQ protocols based on CRC error-detecting codes", *IEEE Trans. Commun.* 43 (1995) 2569–2575.

L. K. Rasmussen and S. B. Wicker, "A comparison of two code combining techniques for equal gain, trellis coded diversity receivers", *IEEE Trans. on Vehicular Tech.* 44 (1995) 291–295.

G. R. Redinbo, "Inequalities between the probability of a subspace and the probabilities of its cosets", *IEEE Trans. Inform. Theory* 19 (1973) 533–36.

C. Retter, "The average binary weight-enumerator for a class of generalized Reed-Solomon codes", *IEEE Trans. Inform. Theory* 37 (1991) 346–349.

C. Retter, "Gaps in the binary weight distribution of Reed-Solomon codes", *IEEE Trans. Inform. Theory* 38 (1992) 1688–1697.

M. Rice and S. Wicker, "Modified majority logic decoding of cyclic codes in hybrid-ARQ systems", *IEEE Trans. Commun.* 40 (1992) 1413–1417.

John Riordan and N.J.A Sloane, "The enumeration of rooted trees by total height" *J. Australian Math. Soc.* 10 (1969) 278–82.

F. Rodier, "On the spectra of the duals of binary BCH codes of designed distance $\delta = 9$", *IEEE Trans. Inform. Theory* 38 (1992) 478–479.

W. Rudin, *Real and Complex Analysis*, McGraw-Hill Book Company, New York, 1970.

K. Sakakibara, R. Iwasa, and Y. Yuba, "On optimal and proper binary codes from irreducible cyclic codes over $GF(2^m)$", *IEICE Trans. on Fundamental of Electronics, Commun. and Computer Science* E82A (1999) 2191–2193.

R. Schoof and M. van der Vlugt, "Hecke operators and the weight distribution of certain codes", *J. Combin. Theory Ser. A* 57 (1991) 163–186.

R. H. Schulz, "A note on check character systems using Latin squares", *Discrete Math.* 97 (1991) 371–375.

R. H. Schulz, "Check character systems over groups and orthogonal Latin squares", *Appl. Algebra in Eng. Commun. and Computing.* 7 (1996) 125–

132.

J. W. Schwartz and J. K. Wolf, "A systematic (12,8) code for correcting single errors and detecting adjacent errors", *IEEE Trans. Computers* 39 (1990) 1403–1404.

R. Segal and R. Ward, "Weight distributions of some irreducible cyclic codes", *Mathematic of Computation* 46 (1986) 341–354.

E. S. Selmer, "Some applied number theory and psychology", *J. Royal Stat. Soc. Series A* 130, no. 2. (1967) 225-231.

N. Seshadri and C.-E. W. Sundberg, "List Viterbi decoding algorithms with applications", *IEEE Trans. Commun.* 42 (1994) 313–323.

C. E. Shannon, "A mathematical theory of communication", *Bell Syst. Tech. J.* 27 (1948) 379–423 & 623–656.

W. M. Sidelnikov, "Weight spectrum of binary Bose-Chaudhuri-Hocquenghem codes" (in Russian), *Prob. Peredachi Inform.* 7, no. 1 (1971) 14–22. [English translation: *Prob. Inform. Transmission* 7 (1971) 11–17].

K. N. Sivarajan, R. McEliece, and H. C. A. van Tilborg, "Burst-error-correcting and detecting codes", *Proc. IEEE Int. Symp. Inform. Theory* (1990) 153.

N. J. A. Sloane and E. R. Berlekamp, "Weight enumerator for second-order Reed-Muller codes", *IEEE Trans. Inform. Theory* 16 (1970) 745–751.

P. Solé, "A limit law on the distance distribution of binary codes", *IEEE Trans. Inform. Theory* 36 (1990) 229–232.

R. C. Sommer, "A note upon another error detecting code that is not proper", *IEEE Trans. Commun.* 33 (1985) 719–721.

D. D. Sullivan, "A fundamental inequality between the probabilities of binary subgroups and codes", *IEEE Trans. Inform. Theory* 13 (1967) 91–94.

L. Swanson and R. McEliece, "Reed-Solomon codes for error detection", *Proc. IEEE Int. Symp. Inform. Theory* (1985) 110.

F. Swarts, A. J. Han Vinck, and H. C. Ferreira, "Undetected error probability of linear block codes on channels with memory", *Proc. IEEE Int. Symp. Inform. Theory* (1993) 162.

W. Szpankowski, "On asymptotics of certain sums arising in coding theory", *IEEE Trans. Inform. Theory* 41 (1995) 2087–2090.

N. Tanaka and S. Kaneki, "A class of codes which detect deception and its application to data security", *Proc. Carnahan Conf. on Crime Countermeasures,* April 6–8, 1977, Kentucky (1977) 221–227.

A. I. Turkin and V. I. Korzhik, "A practical optimal decoding algorithm for arbitrary linear codes having polynomial complexity", Presentation at *1991 IEEE Int. Symp. Inform. Theory.*

N. H. Vaiday and D. K. Pradhan, "A new class of bit- and byte-error control codes", *IEEE Trans. Inform. Theory* 38 (1992) 1617–1623.

J. Vaideeswaran and S. K. Srivatsa, "All unidirectional error-detecting code for universal applications", *Intern. Journal of Systems Science* 24 (1993) 1815–1819.

G. van der Geer and M. van der Vlugt, "Weight distributions for a certain class of codes and maximal curves", *Discrete Math.* 106 (1992) 209–218.

G. van der Geer, R. Schoof, and M. van der Vlugt, "Weight formulas for ternary

Melas codes", *Math. of Computation* 58 (1992) 781–792.

J. H. van Lint, *Introduction to Coding Theory*, Springer-Verlag, New York 1982.

J. Verhoeff, *Error Detecting Decimal Codes*, Mathematical Centre Tract 29, The Mathematical Centre, Amsterdam, 1969.

N. R. Wagner and P. S. Putter, "Error detecting decimal digits", *Commun. ACM* 32 (1989) 106–110.

J. Wakerly, *Error Detcting Codes, Self-Checking Circuits and Applications*, North-Holland Publishing Company, New York, 1978.

X. M. Wang, "Existence of proper binary $(n, 2, w)$ constant weight codes and a conjecture", *Science in China*, Series A (in Chinese) 17, no. 11 (1987) 1225–1232.

X. Wang, "The existence and conjecture of binary $(N, 2, \omega)$ optimal error-detecting constant weight codes", *Scientia Sinica Series A - Mathematical Physical Astronomical & Technical Sciences* 31, iss.5 (1988) 621–629.

X. M. Wang, "The undetected error probability of constant weight codes", *Acta Electronica Sinica* (in Chinese) 17, no. 1 (1989) 8–14.

X. M. Wang, "Further analysis of the performance on the error detection for constant weight codes", *Journal of China Institute of Communication* (in Chinese) 13, no. 4 (1992) 10–17.

X. Wang, "The undetected error probability of binary $(2m, 2, m)$ nonlinear constant weight codes", manuscript 1993.

X. M. Wang and Y. X. Yang, "On the undetected error probability of nonlinear binary constant weight codes", *IEEE Trans. Commun.* 42 (1994) 2390–2393.

X. M. Wang and Y. W. Zhang, "The existence problem of proper constant weight codes", *Acta Electronica Sinica* (in Chinese) 23, no. 7 (1995) 113–114.

R. K. Ward and M. Tabaneh, "Error correction and detection, a geometric approach", *The Computer Journal* 27 no. 3 (1984) 246–253.

R. L. Ward, "Weight enumerators of more irreducible cyclic binary codes", *IEEE Trans. Inform. Theory* 39 (1993) 1701–1709.

G.N. Watson, "Theorems stated by Ramanujan (V): Approximations connected with e^x" *Proc. London Math. Soc* (2) 29 (1929) 293–308.

J. H. Weber, C. de Vroedt, and D. E. Boekee, "On codes correcting/detecting symmetric, unidirectional, and/or asymmetric errors", *Proc. IEEE Int. Symp. Inform. Theory* (1990) 86.

M. N. Wegman and J. L. Carter, "New hash functions and their use in authentication and set equality", *J. Comp. System Sci.* 22 (1981) 265–279.

V. K. Wei, "Generalized Hamming Weights for Linear Codes", *IEEE Trans. Inform. Theory* 37 (1991) 1412–1418.

S. B. Wicker, "Reed-Solomon error control coding for Rayleigh fading channels with feedback", *IEEE Trans. on Vehicular Tech.* 41 (1992) 124–133.

S. B. Wicker and M. Bartz, "Type-II hybrid-ARQ protocols using punctured MDS codes", *IEEE Trans. Commun.* 42 (1994) 1431–1440.

S. B. Wicker and M. Bartz, "The design and implementation of type-I and type-II hybrid-ARQ protocols based on first-order Reed-Muller codes", *IEEE Trans. Commun.* 42 (1994) 979–987.

K. A. Witzke, "Examination of the undetected error probability of linear block codes", M.A.Sc. Thesis, Dep. Elec. Eng., Univ. British Columbia, Vancouver, Canada (1984).

K. A. Witzke and C. Leung, "A comparison of some error detecting CRC code standards", *IEEE Trans. Commun.* 33 (1985) 996–998.

J. K. Wolf and R. D. Blakeney II, "An exact evaluation of the probability of undetected error for certain shortened binary CRC codes", *IEEE MILCOM 88,* (1988) 287–292.

J. K. Wolf and D. Chun, "The single burst error detection performance of binary cyclic codes", *IEEE Trans. Commun.* 42 (1994) 11–13.

J. K. Wolf, A. M. Michelson, and A. H. Levesque, "The determination of the probability of undetected error for linear binary block codes", in *Int. Conf. on Commun.*, Denver, Colorado, June 14–18, 1981, 65.1.1–5.

J. K. Wolf, A. M. Michelson, and A. H. Levesque, "On the probability of undetected error for linear block codes", *IEEE Trans. Commun.* 30 (1982) 317–324.

J. Wolfmann, "Formes quadratiques et codes á deux poids", *C. R. Acad. Sci. Paris Ser. AB* 281 (1975) 533–535.

A. Wyner, "The wire-tap channel", *Bell Syst. Tech. J.* 54 (1975) 1355–1381.

S. T. Xia, F. W. Fu, Y. Jiang, and S. Ling, "The probability of undetected error for binary constant-weight codes", *IEEE Trans. Inform. Theory* 51 (2005) 3364–3373.

S. T. Xia, F. W. Fu, and S. Ling, "A lower bound on the probability of undetected error for binary constant weight codes", *IEEE Trans. Inform. Theory* 52 (2006) 4235–4243.

S. T. Xia, F. W. Fu, and S. Ling, "A lower bound on the probability of undetected error for binary constant weight codes", *Proc. IEEE Int. Symposium on Information Theory* (2006) 302–306.

S. T. Xia and Y. Jiang, "Bounds of undetected error probability for binary constant weight codes" (in Chinese), *Acta Electronica Sinica* 34, no. 5 (2006) 944–946.

D. Xu, "A dual theorem on error detection codes" (in Chinese), *Journal of Nanjing Aeronautical Inst.* 24 No. 6 (1992) 730–735.

V. Yakovlev, V. Korjik, and A. Sinuk, "Key distribution protocol based on noisy channel and error detecting codes", *Lecture Notes in Computer Science* 2052 (2001) 242–250.

H. Yamamoto and K. Itoh, "Viterbi decoding algorithm for convolutional codes with repeat request", *IEEE Trans. Inform. Theory* 26 (1980) 540–547.

Y.X. Yang, "Proof of Wang's conjecture", *Chinese Science Bulletin* (in Chinese), vol. 34, No. 1, pp. 78–80, 1989.

H. Zimmermann and S. Mason, *Electronic Circuits, Signals and Systems*, John Wiley & Sons, Inc., New York, 1960.

V. V. Zyablov, "The performance of the error correcting capability of the iterative and concatenated codes" (in Russian), in *Digital Information Transmission over Channels with Memory*, Nauka, Moscow 1970.

Index